The Bay Shrimpers of Texas

WITHDRAWN

RURAL AMERICA

Hal S. Barron
David L. Brown
Kathleen Neils Conzen
Cornelia Butler Flora
Donald Worster

Series Editors

❖ ❖ ❖ ❖ ❖ ❖ ❖ ❖ ❖ ❖ ❖ ❖

The Bay Shrimpers
of Texas

RURAL FISHERMEN IN A GLOBAL ECONOMY

Robert Lee Maril

UNIVERSITY PRESS OF KANSAS

Published by the University Press of Kansas (Lawrence, Kansas 66049),

which was organized by the Kansas Board of Regents and is operated

and funded by Emporia State University, Fort Hays State University,

Kansas State University, Pittsburg State University,

the University of Kansas, and Wichita State University

Library of Congress Cataloging-in-Publication Data

Maril, Robert Lee.

The bay shrimpers of Texas : rural fishermen in a global economy /

by Robert Lee Maril.

p. cm.—(Rural America)

Includes bibliographical references and index.

ISBN 0-7006-0703-X.—ISBN 0-7006-0704-8 (pbk.)

1. Shrimp industry—Texas. 2. Shrimpers (Persons)—Texas.

3. Competition, International. I. Title. II. Series: Rural America

(Lawrence, Kan.)

HD9472.S63U544 1995

338.3'7253843'09764—dc20   95-4191

British Library Cataloguing in Publication Data is available.

Printed in the United States of America

10 9 8 7 6 5 4 3 2 1

The paper used in this publication meets the minimum requirements of

the American National Standard for Permanence of Paper for

Printed Library Materials Z39.48-1984.

For Andrea, Jordan, Lauren, and Travis

CONTENTS

❖   ❖   ❖

INTRODUCTION

❖ ❖ ❖

Shrimp fishing is the most important commercial fishery in the United States. In the early 1990s American shrimpers docked over 300 million pounds of shrimp per year, valued at more than .5 billion dollars.[1] The Texas shrimp industry is the most productive of all state fisheries. This book is about the men and women who fish for shrimp in the remote bays, rivers, and estuaries of the Texas coast. The work of finding, netting, and docking shrimp is the means by which they have earned their livelihood for many years. Some Texas bay shrimp fishermen can trace their skills in fishing for a living over several generations, but others are relative newcomers. In either case the work they do is often etched in their sunburned and lined faces, in their aching bodies, testimony to lives spent working on the water.

If bay shrimping is a hard life, it is also one that most fishermen and their families say they would sorely miss. Although they indeed rely on the income from their labor—they could not get by without it—bay shrimping sustains them in other ways as well. The work of Texas bay shrimping shapes the lives of these men, women, and children, marks every corner of their existence. It is reflected in the ways in which their households must be arranged to accommodate the demands of the fishing effort; the daily lives of fishermen are shaped by their time on the water.

The men and women who fish for a living do not simply respond passively to their occupational demands and constraints. This book explores in detail the many ways in which Texas bay shrimpers and their families have continuously adapted to the political and economic realities confronting them. Responding to change is not new for bay shrimpers; these men, women, and their families have undergone several distinct periods of transition since fishermen first netted shrimp from Texas bays and rivers.

The Texas bay shrimp industry, like the larger American commercial fishing industry of which it is a part, is witnessing rapid change. One of the most important challenges the commercial fishermen face is strong foreign competition. In the 1990s Texas bay shrimpers compete with Ecuadoran and Chinese peasants who raise shrimp in ponds and enclosed bays and then export them to American shores. Texas bay shrimpers are also at odds with other Americans over the use of the Texas and Gulf of Mexico fisheries. Recreational and sports fishermen have laid claim to fish species that they believe bay shrimpers are systematically destroying in their trawling nets. Moreover, certain environmental groups, with the help of state and federal governments, are at war with shrimpers. Coalitions have been formed, Political Action Committees (PACs) funded, and media campaigns initiated to inform the general public of the harm that commercial shrimping is inflicting on the marine environment. Texas bay shrimpers, in the 1990s, have been branded as poachers.

Increasing conflicts with politically powerful recreational and environmental groups have served to intensify ongoing disputes among Texas shrimpers. The Texas shrimp fishery historically has been divided into two distinct sectors, the bay industry—the subject of this book—and the offshore Gulf of Mexico shrimp fishery, the focus of an earlier study.[2] As they have so many times in their mutual past, the two groups once again find themselves at odds over a limited supply of shrimp.

Since the 1970s Texas bay shrimpers also have struggled with immigrant newcomers to their fishery. Vietnamese shrimpers now constitute an important part of the Texas shrimping industry, both the bay and the offshore fishery. I explore the history of the uneasy relationship between Texas bay shrimpers and their Vietnamese counterparts in chapter 8.

The role of women in the Texas bay shrimp fishery forms an essential part of the industry. Women who shrimp for a living have been interviewed for the first time; their comments on the fishing work they do, their perceptions of their work and their lives on the bay, and the gender-related problems they encounter provide additional insights into Texas bay shrimping.

❖ ❖ ❖

BASIC OBJECTIVES

The first, and major, objective of this book is to document in detail who Texas bay shrimpers are, what they do, and how they do it. Unfortunately, a number of popular myths and misconceptions about the fishermen and their families over time have taken on a life and credibility of their own, independent of any social facts. By providing a detailed description of the lives of the men and women who fish the Texas bays, I have tried to put to rest such distortions.

A second objective is to study the policies and laws that shape the fishermen's time on the water. Of equal import here is an analysis of the motives of the individuals and groups who formulate particular policies and laws to fit their own agendas. I focus on the people who manage and regulate Texas bay fishermen and on the motivations of the interest groups that lobby for laws determining how shrimpers must fish. The interest groups, and the elected and appointed officials who represent them in the public arena, deserve fair and equal study. It is impossible, in short, fully to understand Texas bay shrimpers without including a discussion of the people with whom they compete.

Therefore, I have placed the specific nature of the work that bay shrimpers do within a political and an economic perspective to emphasize the importance of historical relationships between local policies and laws that directly influence the fishermen and the economics of their work on the bay. I have also presented a broader view of regional and national policies and of the interest groups that generate them within a historical perspective and within the context of global economic and political forces.

A political economy of Texas bay shrimpers allows comparisons with other American workers in other industries; although Texas bay shrimpers and the work they do is at first glance relatively unique, there are a number of problems that the fishermen have in common with other American workers. A political and economic perspective provides the opportunity to question the value of the fishermen's work to society, a value that may supersede either a political or an economic contribution.

An examination of the people who fish the bays of Texas and of the political and economic forces that in part form and drive them should resonate among social scientists who study rural peoples and their social and physical environments. Public policymakers and fishery managers will also benefit from the analysis of Texas bay shrimpers and the questions that are raised. But my larger hope is that the general reader, who may live many miles from an American coastline, will have the opportunity to understand and appreciate the complexities of the work and lives of the people who have chosen to fish for a living.

❖ ❖ ❖

METHODOLOGY

I employed both quantitative and qualitative methods to collect and analyze data about Texas bay shrimpers. Following Norman K. Denzin's use of methodological triangulation, I relied on one major quantitative approach, a random sample survey of bay fishermen, and two qualitative approaches, participant-observation and historical analysis.[3] The three methods, when correctly applied, reinforce and complement each other. Additional data were also collected; methods included among others were a purposive survey of women who fish the Texas bays, a purposive survey of Vietnamese shrimpers, and a wage study of Gulf shrimpers.

Quantitative Approach

In 1988 and 1989 I conducted a random sample survey of 154 Texas bay shrimpers along the Texas coast, representing 2.8 percent of the estimated population of 5,453 Texas bay fishermen. The survey instrument consisted of 103 items, of which approximately 80 percent were closed-end, the remainder open-ended; shrimpers were encouraged to expand on their answers, which they sometimes did in great detail.

The majority of Texas bay shrimpers live in small, rural communities along the Texas coast, often some distance from large metropolitan centers. I made every reasonable effort to ensure that the bay shrimpers who were interviewed were representative of the diversity of the total bay shrimpers' population.[4] Fishermen were randomly selected from twenty different locations along the coast, stretching from south of Galveston Bay to Corpus Christi; bay shrimping south of Corpus Christi is of little economic importance. With the exception of Corpus Christi, the locations where shrimpers were interviewed reflected the rural nature of the population.

Rural Texas fishing communities were selected to reflect diversity in race, ethnicity, and socioeconomic status. The survey instrument was translated into local Spanish. Shrimpers were interviewed in old, established fishing communities as well as in the newer, sometimes transient communities of trailers, pickup campers, and mobile homes.

Shrimpers interviewed were selected randomly to decrease the probability of any confounding biases. The fish houses, docks, and other locations they frequented were identified and visited at various times, so that those who fished predominantly in the mornings, or during the afternoon hours, or at night had an equal chance of being interviewed.

Respondents in the sample were asked for basic demographic characteristics, detailed descriptions of the vessels and gear they used, attitudes toward fishing as an occupation and a style of life, and opinions of prevailing policies and regulations governing the fishing of Texas bays. The survey interviews commonly took from twenty minutes to an hour and a half. When possible, shrimpers were interviewed alone so that they would not be influenced by the attitudes of other fishermen, members of their families, or dockside workers. The two interviewers, myself and a research assistant, probed the shrimpers about their responses when appropriate and found that the majority of them were forthcoming in their answers.

Qualitative Approaches

In addition to conducting a random sample survey of bay shrimpers, for six weeks during summer 1988 I was a participant-observer among the men and their families in one small coastal community. Each day I visited a popular dockside cafe frequented by local bay shrimpers. They stopped at the cafe when it opened at eight in the morning, just after docking their boats from a night of fishing in the bay or after an early morning trawl in the nearby river. Later in the morning other bay shrimpers came by the cafe to drink coffee and talk before setting out for their day's fishing or in preparation for the next night's fishing. Initially I met a variety of shrimpers who sat at the rickety tables having breakfast, sipping their morning coffee, relaxing with an open can of beer, or playing pool on a torn and uneven pool table that overlooked a screened porch and the rundown wooden docks. Around one in the afternoon, I usually returned to my car to record in my journal what had transpired.

During the initial participant-observation I met and talked with twenty-six individuals: bay shrimpers, their wives and girlfriends, teenaged sons and daughters, and the staff of the cafe. Every fisherman had shrimped the bays and rivers of Texas on a full- or part-time basis. Discussions centered on their daily concerns, such as mechanical problems with the boats, the amount and content of the catch, and plans for when to fish next. As I became a fixture in the cafe, I turned the talk to general problems of the shrimping industry and to shrimping as a way of life.

Later that summer I spent five nights aboard a bay boat working as the deckhand, a fishing experience preceded by four weeks of being a participant-observer in 1987 on four different bay boats. During both fishing trips in 1987 and 1988 I observed the fishing effort, technique, and strategies of the bay fishermen; at the same time I helped the fishermen with their work, my skills increasing with each trip. By the time of the fishing trips in 1988 I was able to help the captain clean, lower, and retrieve the nets, cull the shrimp on the back of the boat and at the docks, ice it, and carry it to the waiting trucks or fish house. I also manned the wheel when dragging the nets for shrimp but always gave it up when we approached the docks and real piloting skill was required. Over the course of the idle trawling hours of five straight nights on the fishing trips in 1988, one captain served as an excellent informant, providing rich and detailed insights I would otherwise not have gathered. I discussed with him at great length his concerns about the shrimping industry, his family, and his perceptions of his future as a bay fishermen.

When I returned to shore after each of these trips I recorded in my journal everything of importance that I had observed, taking particular care to describe the uses of the fishermen's gear and their fishing strategies. I also paid special attention to the information that the men communicated to each other on their marine CBs while they trawled.

The second qualitative method I used was historical analysis. Very early on it became evident that the attitudes of the fishermen toward their work were grounded in their previous work and life experiences. I became convinced, after concluding my random sample survey and participant-observation, that a complete understanding of the bay shrimpers, and of the contemporary public policies and interest groups against which they struggle, was possible only if the history of the bay industry were also examined. This historical analysis revealed certain patterns in the formulation and enforcement of public policies and regulations and was invaluable in placing contemporary events within a larger, more complex, context.

I first collected the available secondary historical sources on bay shrimping in Texas and nearby states; then I studied articles from several community newspapers as far back as the record would allow. Because of the Texas sesquicentennial celebration, complete records of local newspapers recently had been put on microfiche and were readily available. I also attempted to collect original documents that described or reflected crucial historical events. Texas coastal county and community libraries often proved to be valuable resources for local historical accounts unavailable in major research libraries. I used specific legal cases and records when they pertained to important conflicts between fishermen, to certain segments of the fishing industry, or to political factions in the communities under study. Finally, I reviewed federal and state agency reports that detailed the history of bay shrimping in Texas and along the Gulf Coast. These materials, which are noted in the bibliography, may be useful to other researchers interested in this and related topics.

These three methods—a random sample survey, participant-observation, and historical analysis—and the use of several other secondary sources form the bedrock for this book. My interpretation of the data strongly relies on my previous study and experiences with the Texas shrimping industry. In 1977 I began to interview shrimpers in Port Isabel and Port Brownsville, Texas, who fished the offshore Gulf of Mexico fishery. One year later I expanded the study to include a random sample survey of 152 Gulf shrimpers in three other coastal communities, additional in-depth interviews of approximately 200 individuals close to the industry, and time as a participant-observer aboard a Texas trawler in the Gulf of Mexico. During this two-year project on Gulf shrimpers, I came into constant contact with bay shrimpers, an experience that continued from 1979 to 1981 while I worked on a related project on coastal resources and peoples along the Texas coast.

Because of my familiarity with the Gulf shrimping industry, beginning in 1983 I was frequently hired as a legal consultant in personal injury lawsuits brought by fishing crews against boat owners. In this capacity I represented the shrimp crews as plaintiffs and the boat owners as defendants in county, state, and federal courts. Testifying as an expert witness brought me into close contact with trawler and fleet owners, marine insurance representatives, law enforcement agents, regulatory personnel, and attorneys who specialized in the practice of maritime law. The federal district court in Brownsville, Texas, where I resided, heard many of the important shrimp cases of the 1980s. Through my work as a legal consultant I often had the opportunity to observe firsthand the machinations of the shrimp industry and to discuss informally with commercial fishermen, environmentalists, government representatives, and their attorneys the differing perspectives held on a wide variety of issues that reached the courts.

My long-term contact with and formal study of the Texas shrimping industry have provided a rich data base for interpretation. In particular, the present study has benefited from my knowledge of the industry over three decades, in viewing events from the perspective of long-term political processes and economic trends.

Familiarity with the Texas shrimping industry greatly facilitated my access to, and especially my credibility with, the bay shrimpers' communities. My credibility was essential, because much of the research on bay shrimpers took place when regulatory and policy decisions that directly influenced bay shrimp fishing were being discussed in the Texas state legislature; at the same time, law enforcement agents were cracking down on bay shrimpers who exceeded their catch limits.

I have made every attempt to guard against the bias of friendships I developed with shrimpers, trawler owners, and others associated with the fishing industry in Texas. I have been helped in this effort by a continual dialogue with colleagues from the fields of sociology, anthropology, agricultural economics, history, and the law.

I have protected the privacy of the people I studied or who served as informants by assigning them fictitious names and other personal characteristics that protect their identity. Doing so in no way limits or constrains the findings, analysis, or conclusions of the study. I have named elected or appointed officials, particularly if their names were a part of the public record. Quotations that are not cited were taken directly from the random sample survey or from participant-observation.

❖ ❖ ❖

ACKNOWLEDGMENTS

I am most indebted to Shirley Fiske who helped see the project through from its inception; I owe her a large debt of gratitude for sticking with me through trying times. Lorry King was also instrumental in helping to get the project off the ground. I want to thank Gary Graham as well. Andrea Fisher Maril collected the data on women who shrimped and also authored the chapter on that topic; her work and insights are greatly appreciated. I would also like to thank my colleagues David R. M. White and E. Paul Durrenberger for the freewheeling discussions we had on several different occasions; the study is stronger for their intellectual contributions. The comments of two anonymous reviewers were also helpful. Mary Chipley once again served as my computer assistant, and I cannot thank her enough for her lasting contribution. Michael Jepson was my research assistant for much of the survey interviewing. Pham Dinh Lan served as my Vietnamese interpreter; I learned much from him not only about Vietnamese fishermen but about other topics as well. Jian Guan assisted in the preparation of the bibliography. Daniel Bourbannais contributed the illustration of the bay shrimp boat. I add a special note of thanks to those who served as informants and must remain anonymous. I took the photographs in fall 1994. I would like to thank Dennis McFaddin and Hudson Bates for their help with the photographs and also Mike Fisher for his support, encouragement, and photographic expertise.

This project was funded in part by grants form the Texas A & M University Sea Grant Program, the Mississippi-Alabama Sea Grant Consortium, and the National Science Foundation. Research was also supported by funding from three Oklahoma State University dean's incentive grants. The opinions expressed are my own and no one else's, as are any errors in fact or judgment.

Finally, I would like to thank the shrimpers and their families who spoke straightforwardly about their lives and their work on the water. Because they were willing to share their knowledge and experiences as well as their insights, problems, dreams, and fears, we may better understand who these commercial fishermen are and what they must do to earn their living from the bays of Texas.

1

The Men, the Boats, and the Fish

❖ ❖ ❖ ❖ ❖ ❖ ❖ ❖ ❖ ❖ ❖ ❖ ❖ ❖

The Texas lands are tabletop flat. They stretch toward the horizon, tinctured with molecules of salt that alert the senses, warn them. The air is heavy, leaden, hinting of differences lying just beyond the thick forests of pine, the rough ranchlands, the irrigated fields of sorghum and cotton. There are other flatlands in Texas but none with the giveaway smell riding on a southeasterly wind, none with a promise lurking just a little farther than the eye can see. The airborne salt travels over the fields, past carefully tended herds of brown and white cattle, until the breeze is stopped by a wall of piney woods, its thick floor of decaying matter lying untouched until hunters come searching for deer.

The Texas coastal lands stretch 400 miles along the Gulf of Mexico, a strip running north to southwest, a slow-arching curve from the bayous of the Louisiana state line to the mouth of the Rio Grande.[1] The low-lying swamps to the north give way to tracts of thick pine woods. Around Freeport, south of Houston, the trees are as uniform as golf tees, forming a dank, opaque curtain that lines the highways and back roads, pushed aside only in a few places by small hidden lakes and swamps that dot the area.

These are lands of languid extremes, of hot, thick summer nights and chilling winters when the winds beat down from the northwest.

The temperatures do not fall much below freezing, but the wet chill creeps between buttons and seams and zippers. In between these two seasons comes a brief but flashy fall and a spare spring that quickly gives way to summer. South-southwest, along the sweep of the curve, the forests disappear, replaced by agricultural rectangles resting under thin layers of herbicides and pesticides applied by farm laborers as carefully as frosting to a cake. The winters here are less extreme, the summers longer and hotter. The tin roofs and tar-paper shacks of the pine woods and swamps are replaced by wood-framed farmhouses, their white paint peeling, or by brick ranchhouses with satellite dishes set in small tracts fenced off from the fields of agratechnology. As in the lands to the north, there are few people: teenaged tractor drivers listening to blaring country tunes, grimy gimme caps pushed back on their foreheads, soaking up the air-conditioning in their glass cages atop the diesels. The lone gas station and convenience store sit at the crossroads on landfill bulldozed five feet high to protect the site against the hurricane flooding. It is the immensity of these lands that is impressive, with the vast stretches of sky that at once surround and expose you.

The Texas coastal lands are also the home of the largest collection of petrochemical plants in the world.[2] Refineries and chemical plants erupt from the flat lands, end points for underground pipelines that surface inside the miles of chain-link fences surrounding the complexes. The factories produce vinyl chloride monomers, benzine, chloroform, and other chemicals of the twentieth century, microcities massed in tight packs on either side of the highway where the roads widen and the potholes deepen. The road signs that indicate their presence fail to suggest the scale of the structures; near Point Comfort on Matagorda Bay the plants run for miles, attended only by a few chemical workers who crawl about the tanks and towers in their company uniforms and brightly colored hard hats.

Further south along the curve of coastal lands, boundless fields of cotton and sorghum are interspersed with slow-moving cattle kept from the highway by carefully tended strands of barbed wire and fence post. As the coastal curve deepens and bends north of Corpus Christi, the sorghum and cotton disappear. The roads continue straight and angled, as if the highway engineers from Austin have tried to impose order upon the broad curve of the land where there can be none. Gnarled oak and mesquite trees bend hard to buck the wind, their trunks rubbed raw in places by cattle.

Corpus Christi, the body of Christ, lies at the deepest part of the Texas coastal bend, an expansive city built upon sand, where full-grown trees reach no higher than a man's head. Corpus Christi is a miniature version of unzoned Houston, 300 miles to the northeast. Between the two cities small Texas towns, communities, and villages—agricultural markets and crossroads drawing their economic sustenance from cotton, sorghum, cattle, piney woods, and the refineries—are scattered about, undifferentiated from hundreds of other Texas towns.

But there are other, different kinds of settlements that dot the vast stretches of Texas coastal lands, communities governed by the possibilities and the demands of the Gulf of Mexico. People in these towns have little to do with the pine woods, the inbred foreign cattle, the high-yield sorghum, or the manufacturing of chemicals. They possess a different history and culture from their neighbors. The inlanders look down to till the land, sideways to tend their cattle, up to the horizon to forecast the weather; the people of the Texas coastal rim, commercial fishermen and their families, look east, toward the ocean. They look to the salt marshes and wetlands and tidal breaks and barrier islands, to channel markers and buoys that clang and echo in the salt wind. They find their living and working environment in the changing colors and textures of bay waters, always in flow, in rearrangement, in motion.

The lives of these fishing people are circumscribed by the five major bays along the Texas coastline in addition to a number of secondary bays and river systems.[3] The major bay systems are defined on the east by a narrow, one-to-two-mile strip of barrier islands that buffer the coast from the Gulf sea. Galveston Bay, at the north end of the Texas coastline, is directly adjacent to Houston; then come

Matagorda Bay, San Antonio Bay, and Corpus Christi Bay, the city of the same name covering its western and southern shores. The Laguna Madre, bordered on the south by the Rio Grande, is the longest span of Texas bay but attracts few commercial bay fishermen in the 1990s.

The five major bays are shallow, grass-bottomed, wide spans of water where inlets abound; they are less than three feet deep in many areas and rarely deeper than ten to twelve feet except in the dredged shipping channels. Secondary bays, some quite large, such as Copano Bay, Aransas Bay, West Bay, and Tres Palacios Bay, feed into the five bay systems, providing additional nurseries and rich fishing grounds. Both natural and dredged cuts allow inflow and outflow from the sea that feeds the bays; hurricanes temporarily rearrange cuts, churning up the admixture of fresh- and saltwater from one year to the next.[4]

Small, isolated fishing communities exist along these Texas bays. Each community, regardless of its location and size, is a variation on a theme, a motif quite different from that of the inland towns only a few miles distant. From Houston to Corpus Christi and south to the Mexican border, settlements ring the natural ports and waterfronts, their wooden, rickety docks occasionally giving way to more massive concrete pilings as in Corpus Christi. Fishing sheds and packing houses are perched on thick wooden timbers sunk into the murky port sludge, generations of barnacle shells clinging to their sides. Across the paved or dirt street that borders the small harbors are larger packing houses, sheds, businesses that cater to the commercial fishermen. Mounds of oyster shells are everywhere; crushed shell paves the parking lots and the narrow alleyways between buildings and is mixed unevenly into the sand, dirt, and gravel roads, where it wedges into tire treads and sticks to the undercarriages of the pickup trucks.

In a few of these communities gentrification has firmly taken hold as inland tourists have been lured to the beaches of brown sand on the barrier islands, to the sports fishing, and to other saltwater recreation. In some fishing communities restaurants, boutiques, T-

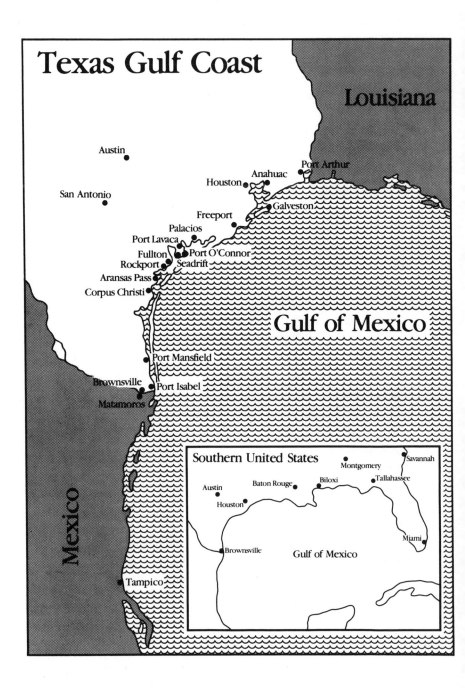

Texas Gulf Coast

Louisiana

Austin

San Antonio

Anahuac

Port Arthur

Houston

Galveston

Freeport

Palacios

Port Lavaca

Port O'Connor

Fullton

Rockport

Seadrift

Aransas Pass

Corpus Christi

Gulf of Mexico

Port Mansfield

Brownsville

Port Isabel

Matamoros

Mexico

Tampico

Southern United States

Montgomery

Savannah

Austin

Baton Rouge

Biloxi

Tallahassee

Houston

Brownsville

Gulf of Mexico

Miami

shirt and shell shops have replaced welding and net shops, marine wholesalers, and diesel-engine repair garages. There is a small row of galleries in Rockport, each with its paintings of local shrimp boats sailing in a Gulf sea, riggings outlined against setting suns, seagulls flying.

The Texas towns began as ports, gateways to the Texas interior. In the communities, most no larger than a few thousand and some much smaller, a downtown business section encircles or extends from the small harbor, depending on the geography of the bay and estuarine system. Some downtowns directly face the bays and barrier islands, some lie safe within natural harbors, a few are at the mouths of rivers or tributaries, others directly exposed to the elements, with the buildings perched high on wooden stilts.

The names of the communities often reflect their geography: River's End, Seadrift, The Lane, Rockport, Bay City, Jones Creek, Ingleside, Aransas Pass, Point Comfort, Portland, Collegeport. A few others, such as Anahuac, were named for the Indians who lived along the Texas coast long before Spanish exploration and settlement.[5] The most notable of these peoples were the Karankawa, a loose affiliation of coastal tribes that did not succumb readily to European settlers. Other communities take their names from their Spanish forebears: Palacios, Matagorda, and Port Lavaca; these ports were often forts, outposts of colonialism, bases for the first excursions into the lands of the Indians.

In Freeport the waterways and channels lie to the north of the original businesses and shops; new businesses, including strip shopping malls, have spread in other directions. In Palacios, south of Freeport, the port area is on the east side of the small town, the business area set back from the shoreline. Downtown Rockport, north of Corpus Christi, is bordered by a small harbor on the east, but the spread of tourist motels and winter residences covers the shoreline to the north and to the south of the commercial docks all the way to Fulton Beach. In very small communities, such as Port O'Connor or River's End, almost all commercial and social life is centered on or near the water's edge as it once was in the larger towns. Businesses are sometimes run from people's houses; the lines between work and home life are indistinct. Bait houses, marine repair shops, fishing-guide businesses, and cafes are often within or extensions of the living rooms and kitchens of their owners.

It is easy to find the fishing harbor in any of the Texas coastal towns by looking for the riggings of the boats. They stand taller than most buildings, which with any height would fall easy victims to hurricane winds and floodwaters. The men on the docks pay little attention to visitors; they are not unfriendly, just intent on other things. They stand in groups of two or three talking, smoking, waiting to work. They look down and out across the harbor, a constant eye on the next shrimp boat that might dock. When a shrimp boat ties up, the dockhands swing into practiced motion and routine; a few more may emerge from the open packing sheds that ring the harbor. One man helps tie the boat lines to the massive dock timbers, joking with the captain as he maneuvers his boat into position. His boat tied up safely at the dock, the bay shrimp captain shuts off the engine and climbs onto the wooden dock. He appears indistinguishable from the dockhands but at the same time different. He wants his boat serviced immediately because he has fishing to do, or he wants to get home quickly to get some lunch before going out on the water again, or he's been out all night and can almost taste a hot breakfast; then he wants to sleep deeply for six hours. He needs fuel and ice for his trip out on the water, or he wants to unload his catch immediately. He wants to know today's shrimp price and how much the dockhands want for gas. He is never pleased with the price. He wants service and at the same time he wants to talk, if only for a few minutes, about the weather, about what the knuckleheads are doing over at the county courthouse, about why shrimping just ain't what it was ten years ago. If the fish house owner doesn't readily appear, the fisherman strides down the dock to root him out, a touch of impatience in the way he walks; he tells the dockhand as he passes to be careful not to overfill the tank the way he did two years ago last fall.

❖ ❖ ❖

THE FISHERMEN

Texas bay fishermen work in the open, and their clothes reflect the nature of their work.[6] They wear white, high-topped, lightweight rubber boots with removable insoles and deep-grooved soles and heels that will keep them from falling on wet decks. The boots are manufactured by a variety of companies, but Red Ball is one of the favorites. The men tuck their pant legs into their boots to keep fish or small crabs from falling in; they wear socks only during the colder months.

The fishermen wear clothes that protect them from the direct rays of the sun and its reflections off the bay waters and from the chill of the water when the sun sets on cold fall and spring nights. Their clothes help prevent burns when they bump against the boat's engine, and their outfits soak up the grease, oil, and fumes common on any work boat. The fishermen's clothes stink of dead fish; the denim and cotton are deeply stained, with marks no washing machine could ever remove. If he wears gloves for protection, his hands still smell of living, dead, and rotting shrimp, fish, and crabs. Indeed his whole body smells of fish even if he meticulously washes his hands after he picks up his nets on each drag; others simply have long since accepted the smell.

Because they spend long periods of time on the water, much of it standing or sitting behind the wheel, the fishermen often move stiffly about their boats and on the dock. Their work requires lifting heavy objects, pulling and wrenching rigging about, working with the winches and boat engines that often do not cooperate, physical work that can take its toll. It is a life of small injuries, most unimportant, some nagging, a few serious. The toll often can be gauged by the men's movements: their limbs and backs may be sore from a recent twist or turn or fall. When the men begin their work, they do so stiffly, awkwardly, until their bodies warm to the task and the minor pain recedes. Their fingers, knuckles, and joints may have been bruised, burned, occasionally broken. Their arms and upper bodies are strong and well conditioned except among the oldest; they carry less weight around their middles than the dockside workers, of necessity being more fit.

The fishermen, both young and old, show the demands of their work on their faces. The long-term exposure to sun and wind and saltwater leaves its mark in the lines running across their foreheads in deep grooves and from their cheekbones to their chins. The backs of their necks are as leathery as any farmer who has spent thousands of hours plowing his fields. Their eyes are often bloodshot, testimony to a night's work. The men often look five to ten years older than they are, or even older. Yet they do not seem beaten down or oppressed; the smiles come quickly, the slap on the back or friendly shove or punch not far behind. And when they speak, it is with a nasal Texas twang, words drawn out but not so much so that you would mistake them for Georgians or Alabamians. The fishermen reveal themselves in their speech as literate but not well educated, as bright, alert, often cynical about what they have seen and done. Almost all of them have opinions about their work and their lives, and they want to be heard. They are not shy or subservient men who are afraid of speaking out or of losing their jobs for doing so. They are men who work with their hands, blue-collar workers, but they also often own their own boats, their businesses. They are direct, to the point; you know where you stand with them because they tell you.

Their demographic characteristics, based upon the random sample survey, provide a profile of the labor force.[7] Texas bay shrimpers are predominantly white, an aging labor force composed of middle-aged and older men, married (72 percent) and with families (the average shrimper has 2.7 children). Despite their many years of experience on the bay, they have limited formal educations. Generalizations must be carefully drawn since bay shrimpers are a diverse work force. White bay shrimpers composed 68 percent of the sample estimate, Mexican Americans 15 percent, Vietnamese 10 percent, and African Americans 7 percent.[8] Bay shrimpers had a mean average age of 42.4 years; about one-third of the fishermen were fifty or older.[9] The average shrimper in the sample had less than eleven years of formal education; about one-fifth had taken college courses.[10]

As measured by the demographic characteristics, Texas bay shrimpers differ from their counterparts who fish the waters of the Gulf of Mexico, who had greater percentages of Mexican Americans, blacks, Mexican nationals, and Central Americans.[11] Gulf shrimpers had larger families on the average, about one child more per family than bay shrimpers; the difference in family size probably reflects the fact that nonwhites generally have more children than do whites. Bay shrimpers were considerably older than Gulf shrimpers; about 40 percent of the latter were in their twenties, and a Gulf shrimper rarely continued to fish after the age of fifty. Bay shrimpers, although not well educated as a group, had more years of schooling than their Gulf counterparts; for example, about 30 percent of the Gulf fishermen had six years or less of school.

❖ ❖ ❖

THE BOATS

Texas bay shrimpers use a wide variety of boats, which can be seen crowding against the pilings of the small harbors of the Texas fishing ports from Port Arthur to Port Isabel. In some communities more than 150 bay shrimp boats line the docks, rafted together, fishermen climbing from one boat deck to the next to reach the dock, their coolers filled with fresh shrimp. Yet in other ports, lone fishing boats pull against their dock lines, the silence of the small empty harbors occasionally interrupted by a passing barge on the intercoastal canal.

The boats come in all sizes, configurations, and conditions, but the fishermen care for them as best they can. If a boat is not in running order, a bay fisherman cannot properly do his work, regardless of his skills on the water. The boats reflect (and are considered to do so by the fishermen themselves) the personalities, expertise, and financial status of the men who fish them.

In an increasing number of Texas ports bay boats are crucial to the holiday ambience. Tourists stare, oblivious of their intrusion, at the bay shrimp boats sitting amid the port sludge and grime. They stroll along the docks at Rockport, for instance, or along the much larger T-heads in Corpus Christi, gazing at the squawking sea gulls and the bay waters lapping against the pilings. And as they stroll, they sometimes stop to buy shrimp from the fisherman.[12]

Bay boats are not the stuff of romantic legends about old men who fish the sea; they are work boats that belch diesel fumes and stink of rotting fish. Dead saltwater catfish float among other scum on the rising tide, and silver-dollar-sized crabs work busily beneath the surface. From the dock children fish with a line and hook for the crabs, shouting when they hook one, crying out in dismay when the crabs escape at the last minute and splash into the water. The docks rock from the weight of the shrimp boats as gunnels push against dock timbers. Barnacles on the beams scrape against the hulls at their waterlines, sounding like the cracking of teacups. The exposed steel strands that strengthen the bulkheads where the saltwater has eaten away the concrete have turned a dark reddish brown; many are deeply pitted, their strength undermined. Silver metal sheds and painted brick and board fish houses in various stages of decay sit next to newer fishing sheds (almost always a sign that a hurricane has passed through) painted sky blue, beige, or off-white.

Even a veteran boat lover must look hard to find a boat worthy of praise. As a fleet, the brightly scrubbed colors of red, blue, white, green, and an occasional yellow stand out against an otherwise drab background, but a closer inspection of each boat quickly belies the first impression. Men in stained cutoffs and dirty white boots climb around a ragtag array of boats. The smallest boats in the survey sample are eighteen-foot flat-bottomed affairs that take on water every time a gentle bay wave comes their way (I saw smaller boats at the dock, but none fell into the sample).[13] At the other extreme are sixty-foot vessels riding five feet higher at the bow than the eighteen-footers. The largest boat in the sample was sixty-one feet in length, a vessel that can regularly fish in the Gulf but that can also drag the rivers and the deeper bays.[14]

The mean length of the boats is 40.4 feet, but because the figure is based upon aggregate data that includes the larger boats, it may give the misleading impression that bay boats are sizable, able to weather the quirky waters of the bays and the Gulf beyond the barrier islands. Generally the boats are not large. Moreover, they are fragile looking, beaten up, old, marine paint falling off in chunks so that any report of bad weather sends most of the fleets of bay shrimpers back to port to ride out the storm.

The majority of the boat hulls (57.3 percent of the sample from the bay fleet) are constructed from a combination of wood and fiberglass. Some fishermen consider these materials to be better than either wooden hulls (16 percent) or fiberglass (19.3 percent). Only 7.3 percent of the boats sampled are steel-hulled, preferred by many because of longevity; then again, some fishermen believe that the initial cost and the extra fuel required to push a steel-hulled boat through the water makes it less desirable.

Many of the men, almost a third of the sample, had built the boats they fished, but the number does not include fishermen who build or help to build boats they sell to other fisherman. Nor does it include men who build boats operated by other bay captains at the time of the interviews. The fishermen in the survey built the smaller bay boats, those less than thirty-six feet; the larger boats are built in the shipyards.

The bay fishermen who build their boats do so according to their own needs and fancies, and those who buy them often make significant structural changes, overhauling the rigging to their own specifications or converting a recreational boat to a work boat. The bay shrimp boats are rigged in a variety of configurations, personalized jerry-rigging being the order of the day. There is no factory that manufactures bay shrimp boats, no dealership where a shrimper can purchase the latest model; the fishermen constantly adapt and build their boats to suit their own needs.

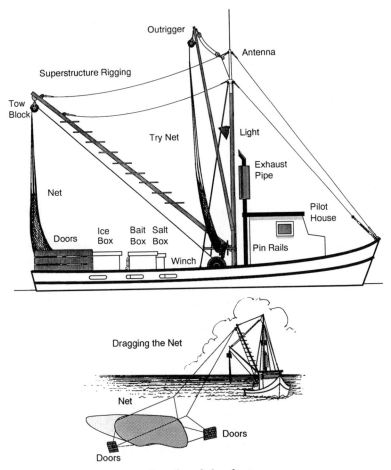

Texas bay shrimp boat.

A small pilothouse sits forward on most bay boats, the major rigging and work deck aft. The size of the pilothouse varies widely: on the larger boats it is a small room with narrow bunks for the crew, a complete galley with stove, refrigerator, and counter space for cooking, and conveniences such as VCRs and stereos. Two of the boats in the sample were air-conditioned. The captain on one of the larger bay shrimpers keeps his broken bicycle against the back of the wheelhouse, intending to repair it when he has some spare time; the bike has been there for two years.

Navigational gear is located on a narrow panel next to the wheel in the pilothouse, along with the marine CB and other communication equipment. The smallest bay boats, often conversions of recreational gasoline-driven boats, do not have wheelhouses; captains stand in open cockpits, exposed to the elements. On some boats the pilothouse consists simply of a wooden roof to cover the wheel; thus sheltered, the captain sits on a rusting lawn chair with plastic webbing, often with an old foam carseat cushion to make it more comfortable. A crude windshield, wooden framed and poorly sealed, protects him from wind and water but is of little use when the rains come or the temperature drops. On the small boats, compass and depth finder are roughly fitted to a wooden panel in front of the wheel. More than once I observed the marine CB hanging precariously by exposed wires to a rotting wooden panel, its marine paint or shellac peeling in thin strips.

Loran, radar mounts, antennas, freshwater tanks, an assortment of lights (almost always including a hand-guided spotlight), and thickly insulated exhaust pipes extending from the engine room sit on top of the pilothouse. Smaller boats have no engine room, the marine diesel being well aft of midship. On the larger boats there are two entrances to the engine room, one fore, one aft; most often the aft entrance is to the port side, directly behind the wheelhouse, the other a hatch that sits a few feet forward of the bow anchor.

The bay boats in the sample are powered by a variety of engines, about a third using General Motors model-671 diesels; four boats that were converted from pleasure crafts to work boats used gasoline engines. On two of these boats the captains say that when they get the money they will replace the gas-powered engines with diesels, which are not only more fuel-efficient but also longer wearing. Yet the captain of the only twin-hull bay boat in the sample originally purchased a pleasure boat, a barge, on which he mounted two gasoline outboards; he believes these engines rather than diesels are best for his boat and does not plan to replace them anytime soon.

The winch or winches are directly behind the pilothouse or the wheel, firmly bolted to the deck. There is considerable variation in the number, size, and condition of the winches in the sample of bay boats. On the larger boats usually three winches are set in a line; the one on the port side lowers and raises the doors of the net, heavy wooden sleds that keep the net's mouth open for the shrimp while the net is dragged beneath the surface of the water. The much smaller starboard winch operates the try net, a small net that tests the number of shrimp present. The middle winch is an all-purpose piece of machinery that raises and lowers the main outrigger and the major fishing net, recaptures the net after dragging, and performs other tasks. A fisherman operates the three winches either by depressing heavy levers with his hand or by pressing on metal pedals with his foot.

The smaller boats have only one winch, which is used for tasks that require the lifting of heavy rigging, lines, and any other gear. The winch lowers and raises the small outrigger, sometimes a jerry-rigged structure with blocks and lines groaning from the weight of the catch as it swings over the side. Smaller boats rarely have try nets. The superstructure rigging also varies greatly from boat to boat. The most common type is a single metal beam amidship, to which is attached the outrigger for the main net and a much smaller outrigger for the try net, if there is one. The main outrigger is lowered into the water by means of a cable attached to one of the winches.

Bay boats generally haul a single net through the water from an outrigger that hangs over the stern of the boat; the size of the net depends on the boat and the season. The single-trawl method differs sharply from the twin-trawl system of the larger Gulf shrimp boats, which fish four nets from the outriggers, two nets on either side.

Steel structural supports are attached to the sides of the bay boats, a crossbeam anchoring the main beam of the superstructure, which is often in the shape of an H although there is much variation. A few of the bay boats (the largest) are rigged to haul a twin trawl, like their Gulf counterparts. They have port and starboard

A boat captain works the winch as he lowers his doors, net, and other gear into the bay.

outriggers that are used in the bays and directly off the barrier islands in the Gulf of Mexico.

A powerful light atop the main beam of the superstructure on the larger bay boats illuminates the work deck, but the smaller boats have a single lone bulb, making it difficult to see the catch and the fishing gear. Three large wood and fiberglass boxes, about four-by-four-by-four feet, stand next to each other on the work deck directly behind the superstructure. One of the three serves as a well where bait shrimp are kept alive, as required by law; a waterhose runs into it and an overflow hose leads from its top to a hole in the gunnel. The second box is filled with fresh water, to which bags of salt and a preservative are regularly added before the boat leaves the dock for a day or a night trip. Shrimp are dipped into the salt box, as some shrimpers call it, before being stored in the third box, which contains crushed ice and serves as a freezer for the freshly netted catch.

The smaller bay boats often have only one box on board; it always serves as the live well for bait shrimp. Other shrimp are stored in large plastic ice chests that the captain shifts about as necessary on the rear deck. Shrimp netted on the small boats are not dipped in salt water, but a preservative is sometimes added to the ice chests in a makeshift fashion.

Racks to store extra wooden doors are affixed to the stern of the larger bay boats; smaller boats have neither racks nor extra doors. Pin rails line the port and starboard gunnels opposite the winches on the larger boats, to which lines are easily tied.

The average bay boat in the sample has modest electronics. Almost all the boats, 95.3 percent, use marine CBs to communicate with each other and with the shore. Depth finders are standard in the sample, appearing on 85.3 percent of the bay boats, as are UHF radios, at 82.9 percent. Loran, an expensive and sophisticated navigational system that almost all Gulf shrimp boats carry, is lacking on 53.3 percent of the bay boats, and only about one in five bay boats has radar. Without it most commercial fishing captains will not fish in the Gulf, even within sight of land, for fear of running into other boats, rock jetties, or other obstructions.

The boats clearly represent various degrees of seaworthiness. Some shrimpers are better boat builders than others; some have more money to spend, selecting better materials and outfitting their boats with better navigational aids. On a number of boats rust has eaten into the rigging and the machinery, marine paint peels from the walls of pilothouses and the hulls, and signs of disrepair suggest neglect. It takes little time for saltwater to take the shine from a working vessel; constant daily maintenance is required on shrimp boats to stem the rot and rust. On working boats, where cosmetic appearance always takes second place to performance, there is a tendency to let some repairs slide; even so, the poor condition of many bay boats in the fleet, both in the sample and among the hundreds of others I observed, far exceeds the look of temporary neglect. The seaworthiness of many of the boats appears questionable at best. Quick and admittedly superficial checks of the engine rooms on some of the bay boats reveal any number of mechanical and electrical hazards, accidents waiting to happen. Vital gear, such as winches, are pitted with rust, wooden transoms rotted through, and on-the-spot, temporary repairs of vital components necessary to the running and fishing of the boat are common. Safety gear, as prescribed by the U.S. Coast Guard, is often in bad shape, hard to reach, or not present. I saw very few bay boats with safety bars on the winch to protect fishermen from getting their limbs or parts of their clothing entangled in the cables.

At the other extreme, some bay shrimpers who were interviewed owned boats in excellent condition, both old and new, including a range of sizes in the bay fleet and representing hours of labor. Such boats are the exception, however; most are often in need of repair, and electrical, mechanical, and structural problems are quite common.

❖ ❖ ❖

THE FISH

The Texas bay shrimp industry is based on white shrimp *(Penaeus setiferus)* and to a much lesser extent on brown shrimp *(Penaeus azte-*

cus).[15] The white shrimp are actually not white, but clear-colored, their exoskeletons transparent; the brown shrimp are slightly darker in color, with a tinge of red on their shells. Although they differ in color, the taste is the same.

Texas bay shrimpers may dehead their catch before they are sold at the dock because intact shrimp spoil faster than those that have "lost their hats." If the bay fisherman catches too many shrimp to dehead before he reaches the dock, he may sell them "with their hats on" to the fish house operator. Many Texans, in contrast to residents of other Gulf states such as Alabama and Louisiana, see only the tail portion of the shrimp at the fish counter or in the grocery store or in the shrimp cocktail at a local restaurant. Detailed drawings of shrimp on menus, restaurant signs, and billboards often depict their anatomy inaccurately; along the Texas coast they are frequently drawn as crustaceans that resemble small lobsters, complete with claws.[16]

Penaeus setiferus and *Penaeus aztecus* in fact have tiny sets of feet that allow them to inch along the bays and river bottoms; for rapid locomotion they use their tails. They have long feelers that extend from small heads dominated by two protruding gray eyes. Their thin, narrow shell bodies are transparent; through their exoskeletons can be seen their simple internal organs. One-third of their bodies is head and thorax, the rest is tail. A thin, sharp barb protrudes from the shell above their heads, which can easily prick the fisherman's hand even if he wears protective gloves. Fingers can swell painfully from too many pricks, a condition sometimes referred to as shrimp poisoning.

Both the white and the brown species of shrimp are unpredictable in their habits. Moving about together in large numbers, they have miles and miles of bay bottom in which to feed and roam so that locating them can be both difficult and time consuming. David R. M. White has observed that the work of Alabama shrimpers is largely determined and constrained by the habits of the fish they hunt.[17] The work of Texas bay shrimpers is similarly shaped by the habits of white and brown shrimp.

The shrimp spawn in the waters of the Gulf of Mexico, and the larvae migrate into the bays and estuaries of the Texas coast; the coastal wetlands, and its adjoining bays and river systems, serve as the nurseries for the five larval stages. Once the shrimp reach their juvenile stage, they move through the bays into the Gulf and attain their full adult size. The exact timing of their movements varies greatly from one year to the next.

Shrimp are bottom feeders. They hide in the mud, emerging to eat small plant and animal life, such as fish larvae, organic trash, and inorganic substances such as sand and mud.[18] They are found together in large numbers, thousands upon thousands covering relatively small patches of the bay or ocean bottom. When the shrimp migrate, they move relatively quickly through rivers, bays, natural cuts, and channels to the open sea. Feeding in the bays before they migrate en masse, they move about sporadically from one place to the next, disappearing from one place only to reappear in perhaps greater numbers in some other part of the bay. Upon reaching the Gulf, they move into deeper waters and begin a generally southern migration down the coast of Texas. The job of finding and netting them in Gulf waters becomes considerably more difficult than it is in the enclosed, shallow bays. Some of the shrimp eventually reach Mexican waters on the other side of the Rio Grande.

White shrimp generally emerge from the mud during daylight hours and are netted by bay fishermen in the mornings, late afternoons, and evenings although sometimes they are active at night. Brown shrimp, in contrast, are usually nocturnal, first beginning to appear as the sun sets; however, I fished on boats that netted brown shrimp in the morning hours before eleven.

Texas bay shrimpers rely on white shrimp for the majority of their catch, and Gulf shrimpers traditionally have fished for brown shrimp in the deeper waters of the Gulf of Mexico. Gulf shrimpers occasionally net large white shrimp because they bring a higher price, on the average, than do the brown ones, but the practice is rare. White shrimp can be netted at different times of the year in the bays and estuaries of Texas as well as in the rivers and cuts that lead to the Gulf of Mexico. A few bay shrimpers fish for white shrimp directly off the barrier islands in the shallow waters of the Gulf of Mexico, netting for brown shrimp as well. When neither is in abundance, bay shrimpers fish other species, including red and popcorn shrimp, which are smaller and bring a much lower price at the dock.

2

Working on the Water

❖ ❖ ❖ ❖ ❖ ❖ ❖ ❖ ❖ ❖ ❖ ❖ ❖ ❖

At first light, the sun still not above the haze that hugs the horizon, Larry greets me with a grunt and a wave. The *Queen Bea's* diesel growls as I stand on the rickety dock, its engine warming slowly in the summer morning, the vibrations rocking the boat, sending small ripples into the shadows of the dark waters of Port Lavaca. A nearby gull perches on one leg on the stub of a rotting timber, eyes closed, wings folded tight against its body. Bird droppings pattern the dock, thin splatterings of white and gray covered with a film of salty dew not yet dried by the rising sun.

Larry motions me aboard with another wave of his hand. He has just finished mending his try net, webbing clutched in his left hand, a sharp-bladed three-inch knife in his right. He sticks the knife in the rear pocket of his stained pants, then quickly ties off the knot that closes the hole. He is a stocky man with long brown hair that he combs straight back; thirty-four, he looks a decade older. Larry is friendly but naturally cautious. This is my fifth trip out with him, and the fishing has made us comfortable with each other, not friends yet, but used to each other's ways.

Two bay boats, one following the other's wake, slowly motor past us in the next ten minutes. "Shit," Larry mutters to no one in particular. "We gotta get out of here before the shrimp are gone."

But we stay at the dock another thirty minutes, time for Larry to finish some necessary boat chores left over from the day before. I help him find a wrench he left next to the engine when he'd replaced the oil filter; then he is off down the dock, ignoring the big holes in the planks as if they did not exist, never breaking stride. A few minutes later I see him under the roof of the fish house talking with another man, making a point with his hands. Then Larry disappears into the gray metal building, coming back with his shirttail tucked in, his red greasy gimme cap reset straight, water droplets on his chin. "It's always something," he says as he unhooks the bow line from the dock cleat, nodding that I should do the same with the line at the stern.

Larry ducks into the pilothouse with a practiced motion, shifts the engine into gear, its drone rising an octave in response. He guides us away from the dock, his right hand on the wheel, standing next to me outside the shelter of the pilot house. The *Queen Bea*, under his fingertip control, motors down the channel, Larry waving to the men he recognizes with his free hand, idle dockworkers standing about drinking coffee from styrofoam cups or smoking cigarettes, other bay fishermen moving around their boats trying to get away from the dock as fast as they can but bound to it, as we had been, by one last chore. Larry calls to them by name, the men responding with a nod or a wave before returning to their work.

Gulls fly in and attach themselves to us as if by a cord, ten feet off our stern, heads high, waiting for free food. They are silent in their pursuit, no screams, but I hear the beat of their wings between the chugs and rumbles of the *Queen Bea*'s diesel.

I find a place to sit on the salt box as we pass by the dilapidated buildings, fishing sheds, seawalls, the nameless junk and rubble that line the port. The tide is up and half covers the remains of two piers that jut into the water from the roofless walls of an old shed. Two boat skeletons, both bay shrimpers, lie sunken in the port mud, each stripped of fittings or anything of value; their hulls look slick as eels, covered with slime. Barely visible through the glassless portholes, a school of beer cans floats in the water that fills the pilot-houses.

The maze of the small port makes sense only from the water. From the land the backs of the flimsy buildings face the parking lots, pocked with deep holes and covered with bottles and weeds. Surrounded by rusting chain-link fence, the piers and docks extend from earth mixed with fill rock, topped by more weeds. But from the *Queen Bea* it is clear that the port has been constructed facing the bay to serve the fishermen. Some of the fish houses stretch down to the water with only a seawall between the bay boat and the roof of the shed. The diesel pumps are handy, off to the side on the dock so fishermen can first unload their catch, then gas up. Some of the fish houses have small weighing scales, their surfaces washed shining clean. Waterhoses are wrapped and put away. Everywhere cracked concrete floor and loose gravel cover the bare ground, the shifting layers of sand and fill beneath raising and lowering the hardened surfaces; weeds grow in the crevices. The port is set up, designed, for the shrimpers: easy in, easy out, then back to the bay to work the shrimp.

We motor down one channel, turn a corner, skirt the rotting wooden hull of another bay boat half submerged in the port sludge. Larry reads my thoughts. "There aren't any junkyards for shrimp boats," he tells me, chuckling to himself at the obvious.

Then we turn a corner and head out into the bay, eastward. With that one turn we enter another world. The wind picks up a notch; the gulls at our stern suddenly regain their voices. Larry throttles up the engine to ten knots, and vibrations from below shake the boat hard, bumping me off the salt box as the boat plows its way through the bay waters. The *Queen Bea* was built to graze the waters of the bay with fishing nets, not to pull skiers, not to run at eighteen knots until she reaches the blue waters off the barrier islands with a load of marlin fishermen; ten knots is close to her top speed. Landing on my feet, I look around the rear starboard corner of the small pilot-house and see a wide mass of gray bay waters stretching to the horizon where the thin dark line of Matagorda Island separates the bay from the Gulf of Mexico. The rising sun backlights the strip of barrier island, making it impossible to see any detail on the narrow wedge, the palm trees, the houses set up high on stilts, the dunes.

The *Queen Bea* staggers gracelessly through the light chop, parting the grayness, exposing white curls and froth, laying wide patterns behind us in our wake that in the absence of other boats continues until lost from sight. Larry's newly patched try net swings back and forth from its small outrigger. The large fishing net makes big, lazy arches in the morning air as it hangs from the aft outrigger that creaks in the light chop, pieces of dead fish breaking loose from its webbing and falling soundlessly to the wooden deck.

Far off in the gray bay waters of Tres Palacios, due north of our position, I see other boats with their nets already in the water, slowly dragging the bottom for shrimp. Their movement is not discernible; they appear stationary in the morning bay waters. They are like distant tractors in a million-acre sorghum field, the details of the boats and their crews too distant to perceive.

Between these boats and the *Queen Bea* is an expanse of water that gradually lightens under the rising run so that the space between us seemingly grows wider. The shades of gray continue to change until quite suddenly the water is a dark, molten green, as if a planet of algae lies beneath our hull. The boat wrestles through the ever-changing green water, her diesel powering us through the bay. The sky turns dark blue and seems to hang so low over our heads that it could be reached and carefully examined. The gulls hammer the boat, nasty about the delay in their breakfast, as diesel fumes thicker than Houston's best smog rumble from the twin exhaust pipes on the pilothouse.

And what of the salt on the wind, the same substance that hangs on the breeze fifty miles inland, that sneaks its way into your lungs, that tinges the wings of the inland birds? Pervading every part of the bay, it plays third-fiddle as Larry guides us to his fishing grounds. There is a vacuum on Matagorda Bay, a huge absence. This bay, to the north Tres Palacios, to the south Espiritu Santu, is totally empty; the vastness is a shock, and I look instead at our little boat.

There is nothing on the bay or the bays that feed into it but our small boat, a few other small boats, the channel markers and buoys, car tires, and some garbage, including the ever-present beer cans, to remind us that people regularly pass by. Port Lavaca lies diminished, a line of small bumps that rises and falls with each rumble of the *Queen Bea*'s engine. I want four-lane highways, exit signs, billboards, grazing cattle, trees, a hundred references to an everyday environment; instead I see and feel changing water, colors and space in flux, an expanding void.

A clean seawind blows across the bay before crashing ashore. I lean into the pilothouse for warmth and familiar sights; the sun's rays, magnified by the glass of the windshield, heat up the small space. Larry draws hard on a cigarette, a cup of steaming coffee in his hand.

"You want some of this?" I nod. One hand on the wheel, he pours me a cup from the pot heating on the galley stove.

"You gonna get sick on me?" Larry asks.

"No."

There is no cause for seasickness. I've been to sea with Gulf shrimpers, know the feeling in the pit of the stomach when the eyes search the horizon for another boat, for any object, something material, something that does not dance and shimmer, something solid, made by human hands. But on a bay boat land is always within sight, often no more than an hour's run before port. We are on a huge lake, surrounded on all sides by civilization but at the same time walled off from it. I am not going to be seasick, but I keep returning to the pilothouse, a warm, gadget-filled place that provides a familiar comfort when the bay gets too empty, too big, too changeable—especially at night.

Larry ties down the wheel, his cheap form of autopilot, and walks back to one of the winches. He lowers the small outrigger into place by playing out the cable on the winch, then lowers the try net, again with the aid of the winch, onto the rear work deck. He ties a knot with a line from the bottom, or bag, of the small try net that he recently mended, checking once again to make sure the holes in its mesh are patched. Then he straightens out the webbing with his hands; the mesh has dried into stiff clumps that need to be smoothed out.

He returns to the pilothouse, reduces the boat's speed to less than a knot, and goes back to the work deck to feed the try net over the starboard side into the bay waters. The slow pull of the water stretches the net out as he slips it over the low sides of the boat. Larry is careful to keep the net and the cables tied to it well away from the boat's prop; an entanglement can delay his fishing for several hours or more.

The try net is attached to a black metal frame with small cross-members to keep it rigid in the water. Larry told me he had designed the frame and built it himself; he is proud of it. The frame keeps the mouth of the net open as it is dragged along the bay bottom. On other bay boats captains use a set of small doors to achieve the same objective.

Larry lowers the try net into the water until it is in place to the starboard and aft of the *Queen Bea*. He walks back unhurriedly to the pilothouse, brings the engine up to a dragging speed of between two to three knots; the boat effortlessly hauls the little net, not more than ten feet in length, along the bottom of the bay. Every fifteen minutes or so Larry leaves the pilothouse to check the try net. When he finds enough shrimp in it, he will lower the large fishing net into the bay water. Some captains systematically check their try nets every few minutes, others let it go for an hour or more, depending on how they feel about where the shrimp are hiding. Larry has told me before that the try net is helpful but cannot always be trusted.

To retrieve it Larry reduces the speed of his boat and winches the net up. When the bag reaches the surface of the water, he quickly throws out a line with a metal hook at its end, catches the try net, and pulls the bag over the side of the boat without much effort. Larry tells me he prefers a line and hook but has also used a hooker pole, a five-to-six-foot aluminum pole with a large metal hook on the end.

He tugs on the knot of the bag he had carefully tied earlier, and the contents fall to the work deck. We quickly sort through the small catch with the heels of our rubber boots. There are a few medium-sized white shrimp, not many, jumping and wiggling on the deck; we scoot them to the side for the moment.

The captain who believes in his try net counts his shrimp, figures the time it has taken to net them, and decides whether it is wise to fish his large net. Bay captains estimate not only the abundance of shrimp that are present from the catch in their try nets but their size as well since size determines the price at the dock. Larry does not like what he sees.

"We're just fooling around now," he tells me. "We'll start getting serious in another hour or so. If they're here, we'll find them."

I sweep the "trash fish" over the side. The try net picks up everything on the bottom of the bay, and with the shrimp comes a variety of small fish, blue crabs and spider crabs of all sizes, jellyfish, hardheads, sea slugs, and small sharks. The try net and the regular fishing net also capture flounder, whitefish, speckled and brown trout, redfish, and other game fish. Rays of all sizes are netted along with eels and other fish whose names the shrimpers do not know. The shrimp are kept; sometimes the larger flounder, trout, and redfish are put on ice. The try net and fishing net also drag up other objects from the bay bottom: beer cans, plastic bottles, fishing gear, clothing, and rocks.

Larry is a bay shrimper who relies on his try net, and he knows from its contents that he does not need to use a "shooter" this trip. There are few jellyfish at this time of the year, but sometimes the cabbage heads are so numerous that he installs a specially designed small net, a shooter, at the mouth of his large fishing net that filters out the jellyfish. Not all bay shrimpers like to use shooters, some considering them not worth the extra hassle, but Larry believes a properly placed shooter saves him time and energy.

Larry picks up his try net and we motor farther south-southeast over the expanse of the bay. Suddenly he stops the *Queen Bea* dead in the water, the diesel muffled. Without a word, he lowers the main outrigger into place with the help of his largest winch. Then he

A deckhand shakes the net to clean it before dumping it over the side for another drag.

gives his thirty-four feet of fishing net a once-over as it drops to the rear work deck, carefully unfolding it, straightening it out, all the while checking for tears. As he inspects the net, he tosses aside bits and pieces of yesterday's catch that have dried in the folds of the mesh. Satisfied, he ties the bag of his fishing net with the same quick release knot he uses with his try net.

Larry increases the boat's speed to one knot and returns to the work deck where he works the bulk of the net over the starboard side, away from the wind. Next he carefully feeds out the net and other rigging, called an otter trawl, into the bay waters, cautious that the mesh does not foul the prop. With one of his smaller winches he raises the set of doors from their rack, then swings them out over the rear deck at head level and over the water, easing the doors into the bay with the help of the winch.

The pull and weight of the doors spreads the net to its full length and breadth as the trawl swings around the stern of the *Queen Bea*. The cable grows taut as the net sinks below the surface and is pulled through the water. Larry increases the boat's speed to between two and three knots, then holds it there for the duration of the drag.

I join Larry in the wheelhouse, leaning against the doorframe of the starboard wall. We talk about nothing in particular for a while, grow tired of our voices, listen to Larry's favorite country-western station.

After about an hour and a half he says, "Well, I guess it's about time to do her."

First Larry picks up the try net, which he left in the water while he was fishing the big net. He cuts the engine speed, quickly boards the try net, empties the small catch, ignoring it until he has more time, and returns to the pilothouse where he shifts into neutral. The *Queen Bea* drifts slowly to a stop, then gently rolls and pitches in the small waves and light wind.

Larry winches in the cable to the otter trawl, the doors breaking the surface of the bay waters. He raises them a few feet from the tip of the main outrigger, over the rear of the work deck, and drops them lightly into the door rack with a thud. Just before the doors

The doors and fishing net are winched aboard after a drag. The rear work deck of the bay boat has a low gunnel to facilitate the raising and lowering of the gear.

slide into their proper place, he runs over to give them a slight nudge with the heel of his white boot, a push that eases them into a tight fit.

He returns to the main winch where he continues to reel in the steel cable, the wet strands wrapping around the thick drum of the winch. The length of the fishing net draws up along the starboard side of the boat. Larry locks the winch in place and quickly pulls the bag of the net, now filled with catch and weighing several hundred pounds, over the rear work deck. He uses the lazy line, a rope that attaches from the stern of the *Queen Bea* to a place just above the bag of the net. The outrigger groans against the weight of the catch, the metal bending, its tensile strength tested. Grabbing the lazy line, which is attached to the doors, Larry pulls the net close enough to get the whip line, another rope, on it. With the whip line wrapped around one of his smaller winches, he raises the bag of the net off the work deck and positions it exactly where he wants it.

Larry works quickly, automatically going through motions practiced hundreds of times, not thinking about each individual decision that must be made, each tug and pull, but focusing on his objective. He always keeps an eye on the cable line to the winch, alert to the smallest sign of a problem. His objective is to get his catch out of the water as quickly as possible and over the side onto the work deck; he does not want the catch swaying precariously from the outrigger, the fishing net out of the water, the weight of fish and wet mesh pulling against the outrigger. As soon as he lands his catch, he wants to get the shrimp sorted, iced, or into the live well. The longer it takes him to do this, the more likely his shrimp will be damaged.

Larry ignores his catch until he has cleaned the trash fish, mud, and other debris from the bag of the fishing net, noting any tears as he works. He reties the knot in the bag with a few sure movements and lowers the net back into the water for the next drag. He detaches the whip line from the net, tying it off at the stern of the boat. Again he hauls up the doors from their rack and carefully drops them over the side with the winch cable. He resumes dragging speed, ties off the wheel onto the desired course, and rejoins me on the work deck.

The catch is a heap of squirming sea life, death, mud, sea grass, and fish slime. I start to sort through it, throwing out any large fish, rocks, or other garbage. The blue crabs, both large and small, run for any cover they can find. The spider crabs, looking twice as ugly as their name, are less swift, their elongated legs caught in the mess of the catch; they unfold themselves slowly and carefully. This load has more rays than usual; I carefully avoid their barbed tails. The majority of the shrimp, though still alive, lie under layers of dead fish and mud; they will soon die. Individual shrimp catapult themselves several feet into the air, attempting to propel themselves to safety.

Larry begins to shovel the catch into the saltwater well, both of us throwing out by hand the larger trash fish, crabs, and other debris from the bottom of the bay. We handle the crabs, catfish, and rays carefully, but it is tedious work. There is so much trash that it is faster and less tiring to scoot some of it along with my boot heel. For my own amusement I begin to punt crabs over the side of the boat.

The sea gulls that trailed us from port swoop in to steal the dead fish, brazenly ducking under our legs and arms to grab as much as they dare. I find them obnoxious; their cries are deafening. The gulls on the wing to the aft of the work deck do not attempt to steal the catch from us but fight the thieves for their food, snatching fish before it can be digested. The gulls who lost their prizes soon return for more.

As we work Larry keeps one eye on the boat's course, careful to avoid running into any obstructions in the bay. He also watches over the tension of the large fishing net's cable that he is still dragging across the bay's bottom.

In a few minutes many of the small fish, sea worms, eels, and other dead trash fish rise to the top of the water in the well. Larry quickly skims them off and flicks them over the side of the *Queen Bea*. Then he dips into the well with the skimmer, scooping out the shrimp, small fish, crabs, and remaining debris that have sunk to the bottom and dumps them onto the top of the freezer box, a bumpy but flat surface.

I quickly cull through this catch as Larry continues to skim the

A deckhand culls through the catch on the rear deck. The catch is mostly
jellyfish, a few shrimp. His rake leans against his left foot as he throws
shrimp into the laundry basket at his right.

salt box. Soon he joins me, and we work with both hands, throwing aside any remaining trash fish and debris, separating the desirable shrimp from the rest. As soon as we finish culling through one batch, Larry sweeps the shrimp into a plastic laundry basket, then again scoops the contents from the well with his skimmer, dumps it onto the freezer top, and we repeat the process. Ever so slowly white shrimp fill the laundry basket.

It takes about forty-five minutes for the two of us to work through the catch Larry landed with his first drag. He grabs the shrimp netted earlier from the try net and tosses them into the laundry basket. By the end of our work, the basket is a third full, roughly twenty-five pounds of shrimp with their heads still on. Larry cleans the shrimp with fresh water from a hose and shakes the basket, loosening up some of the remaining mud; the water running through the shrimp in the basket grows clearer.

Larry's insulated ice box is divided into two small sections by a wooden panel; on one side is ice. He shovels the ice from one section into the other, pours the shrimp from the basket onto it, and places another layer of ice on top of the shrimp to freeze them and keep them from spoiling. On smaller bay boats, shrimp are often placed in coolers in the same fashion. We had already sorted the few brown shrimp from the white. Larry sticks the largest shrimp, those that will bring the highest price, in the corner of the freezer, away from the rest, along with the brown shrimp. I help him gather the flounder we netted and toss them on some ice, too.

We wash off, dunking our hands up to the elbows in the salt well. Larry tosses me a piece of Lava soap, which helps remove some of the grunge from my hands; the suds bite at the tiny cuts in my fingers and palms. We drink more hot coffee although now the day is warming; it is already eight-thirty. There is a twenty-minute break before Larry stops the *Queen Bea* dead in the water and brings over his second drag of the trip.

"Nothing much happening around here," he tells me, looking at his second catch. "One more drag after this next one, and we call it a day." Then we'll anchor up somewhere quiet and smooth in the bay, clean up the boat, do some repairs, and sleep for a few hours before sunset. Usually Larry motors back to the dock and drives home to see his family before the night fishing, but his wife has taken the kids to see her mother and he wants to get some extra work done on the boat and, he hopes, make some money.

If Larry were bringing a lot of shrimp over the side, we would keep working, sorting through the trash, culling it, cleaning the shrimp, finally icing it. Dragging time depends on how much shrimp we net: the more we net, the shorter the time to drag. Larry keeps his try net in the water, constantly estimating when he should bring up his fishing net.

If a bay shrimper is catching few shrimp, the drags are longer and the work in between them slight. I did not meet any captains who preferred idle time to work. Work makes the time go faster; the shrimper feels better because he knows he is making money. Time on his hands often drives him to his CB. Standing at the wheel, the second drag in the water, Larry surveys the CB channels and selects the one he knows his friends prefer. He listens to a conversation between two boat captains, identifying them for me by pointing the boats out as they lie rocking gently in the distance.

"That's Stan," he tells me.

"How do you know?"

"Because that's just the way he talks. You can pick him out any day. Nobody around here talks the way Stan talks."

Stan talks, to my ears, the same way the other shrimpers on the CB talk. But Larry knows each of them, has fished with them for ten years. In fact, he keeps referring to Captain George in the *Miss Liberty* as the "new guy"; Captain George fished part-time on the bay for twelve years but only began to fish full-time three years ago.

Larry rarely introduces himself on the CB; he just waits for a pause in the conversation and then talks. Usually he talks about the bad price for shrimp, the weather, boat repairs, or some combination of the three. When Larry gets bored, he begs off with "Well, that's about it for me."

Then he tunes in his favorite country-western station, but when the local or national news comes on, he changes channels. He tells me he's sick of hearing the same bad news played over and over. When he turns off the music station, we are left again with the sights and sounds of the bay, the desert view of water constantly in motion and the reassuring throb of the *Queen Bea*'s diesel.

❖ ❖ ❖

NIGHT FISHING

I sleep for two hours on the narrow bunk in the pilothouse. Larry says he is going to, too, but instead he repairs some wiring under the wooden instrument panel. By the time he finishes, it is already 5:00 P.M., the sun beginning its slow descent into Port Lavaca Bay.

I stagger around the rear deck with a cup of coffee, my eyes adjusting to the light, ignoring the stench of my own clothes. There are no other bay boats in sight, but soon some will be motoring out from the docks in Port Lavaca, and others anchored up as we are to wait out the daylight, will eventually fall into our sight. I yawn and stretch as Larry goes about his routines after putting away his tool box. First he checks the nets, lines, and cables, motioning me to clean up a patch on the rear deck I missed; then he goes below to take a look at the engine.

Around 7:00 P.M. the slight wind dies away and the waters flatten, multicolored as the sun's light fades. We sit gently rocking in the shallow bay, the tops of the sea grass visible beneath the bow. I watch the sun as it descends directly behind the town of Port Lavaca, its line of fish houses and a water tower now backlighted by an orange glow. The waters around us change in turn, from green to dark blue to gray in the growing twilight. Larry switches on the running lights as he warms up the *Queen Bea*'s diesel. In the twilight I now can see the other shrimp boats, ten or so faraway objects floating, as we bide our time until the shrimp appear.

Under way again, Larry heads for one of his favorite spots, to me no different from any other place in the bay but to him a location where he feels he is almost guaranteed to catch some shrimp. He lowers the try net into the water, and I join him in the wheelhouse. He points to our location on one of his marine maps.

"Why there?" I ask.

"Because that's where the shrimp are."

"How do you know they're there?" I say, disbelieving.

"They gotta be somewhere," he explains, swallowing a chuckle.

Larry readies his fishing net. He saw the try net break the water as he reeled it in and does not need a second look to know that he must get the big net dragging; the try net was filled with fish, a good portion of them shrimp. We are on top of them now and may soon pass them by. Once Larry has the fishing net in the water, he smiles to himself and gets on the CB to talk to his buddies, the closest a mile away. He does not mention to them the shrimp he knows he is catching in the big net.

Soon, to the northeast of us, the lights at the Alcoa plant come on. Although ten miles away, the lights from the huge plant shine on the chop, laying deep shadows over the shallow waters. The lights divide the *Queen Bea* in two, her starboard side now a haze and blur of indistinct forms and colors, her port dark as the other side of the moon.

I do not like working at night. Even under the bright worklights that hang from the rigging, the work deck is a quilt of distorted shapes and patterns. The trash fish take on weird shapes under the alternate glare and shade, the spider crabs turning into alien forms as they fight free from the mass of the catch. I guess at what is and should be before my hands, confirming only by final touch. Larry is not bothered by the absence of light and the necessary reliance on his other senses.

"It's spooky out there," I tell him, pointing toward the Alcoa plant.

"Nothing but water," he replies. "Some people take to it and others don't." He retreats to the pilothouse.

The wind picks up again about 11:00 P.M. We are fishing for

brownies now. The third drag of the night is in the water, and Larry counts his money in his head; he is in a great mood.

"How much, you think?" I ask him. We have just finished putting the shrimp from the second drag into the freezer box.

"I figure so far about 175, maybe 200 pounds. We keep going that good and I'll make some real money tonight. For a goddamn change."

At 4:00 A.M. we finish culling the catch from the last drag. Larry figures we have a good 500 pounds of shrimp, maybe more. We are already 100 pounds over the limit, but Larry is not worried about running into the game warden.

"They're not here. Haven't been seen for three weeks. If they'll stay away a little bit longer, I can earn a living this year the way I used to."

I sit wearily at the starboard doorway to the pilothouse, my hands sore, my legs stiff and hurting, my eyes heavy. Larry talks again to his buddies on the marine CB, something about the best place to buy boat paint in Victoria.

I drowse off and awake with a start at 5:30 A.M., just before dawn, the eastern sky vaguely light. Larry leans forward hard on the wheel, which he's tied off. I barely see him in the new morning light.

"You ever fall asleep when you're fishing?" I ask him, half testing to see if he is still awake.

"Every once in a while. But not often." He readjusts himself, grabs some matches, and lights a cigarette. If he was not sleeping at the helm, he was close to it.

"I'm usually too scared of running into something to fall asleep," he says, savoring the smoke. "What kept me going just now was thinking about all those shrimp we got in the freezer."

"How'd you know where they were last night?" I ask him, stretching my legs, standing up carefully, my back stiff as a board. I reach for the coffee pot.

"This place has always been good to me before. It was again last night."

My mind clears with the coffee, and my stomach begins to churn; I'm hungry. It occurs to me that Larry knows 100 percent

A boat captain proudly displays the largest shrimp in his catch from the bay.

more about the bottom of this bay and the shrimp that live in and around it than I can ever hope to know; his try net and his long experience outline and detail it for him, giving him the clues he needs to find the shrimp. His true fishing world is the space beneath the waters, the slope of the sand and mud, the abundance of the sea grass that fouls his nets, the ratio of small trash fish to shrimp, all of it. Details insignificant to me are important sources of information for Larry; he makes decisions based on his experience.

Larry's world is also the chemistry of the water itself, its balance of acid and base that keeps the shrimp at home in their burrows or sends them en masse from one side of the bay to the other. And Larry always wants to know the water temperature, weighing that against the other details he knows. He does not care if it is day or night; he does not need light to fish or to keep from falling overboard, like me. Every inch of the *Queen Bea* is a given in his life on the water. He puts the boat through its paces, always determined to find shrimp; his work routines are ingrained in him, a means to an end.

He told me one time that he knew another captain who tasted the bay water with his tongue.

"Does it work?" I asked.

"I don't know. I'm just telling you what he does."

Larry may be ambivalent about water tasting, but he believes in the importance of the moon. He knows its phases whether it is hidden by clouds or bright enough to cast shadows of the *Queen Bea*. Like many of the other bay shrimpers, he believes that the shrimp often move in direct response to the phases of the moon. If a captain is not out on the bay under a summer's full moon, then he is a complete fool and should sell his boat and gear to a real fisherman.

❖ ❖ ❖

THE DECKHAND

Larry holds strong opinions about the usefulness of a deckhand on board the *Queen Bea*. Unlike most bay fishermen, he rarely uses one unless absolutely necessary. If he pulls a muscle in his back and cannot work the fishing net or if the shrimp are so plentiful that he needs an extra hand to help cull, then he will hire a deckhand, but only temporarily.

The vast majority of captains in the survey sample, 87.5 percent, say they regularly employ a deckhand to help them fish for shrimp in the bay. Although no captains in the sample responded that they hired more than one deckhand, I see a few bay boats with more than one; Vietnamese bay boats, in particular, frequently carry more.

The role of the deckhand on a Texas bay boat is to help the captain with all the work that must be done.[1] At the same time, the deckhand serves an equally important social role; he keeps the captain company out on the bay. Larry makes a big point of telling me that he will not use a deckhand unless he absolutely needs him, yet he admits that the advantage of hiring a deckhand is that "it is somebody to pass the time with." Having someone to talk to while performing the often boring and repetitive work makes the fishing easier.

A good deckhand is the captain's second pair of hands. He helps with the equipment and gear, takes his turn at the wheel, culls and heads the shrimp, cleans and maintains the boat. Deckhands are most useful when the fishing net is dropped overboard or picked up after a drag since the captain cannot be at the wheel and on the rear work deck at the same time. Larry solves this problem by tying off his wheel and keeping a careful watch when he is on the work deck. This practice is not safe, however; there is always the possibility that the *Queen Bea* will run onto a spit of sand or some other obstruction. And Larry wastes time and energy running back and forth from the pilothouse to the rear deck.

Deckhands are also vital in an emergency. If the boat's engine dies or if the doors or the net cables are fouled, a deckhand provides the additional help that is required. I observed many deckhands who did not need to be given orders or directions; they always knew what was expected of them, and they did it.

The deckhand on a bay boat generally performs the same duties and tasks as the captain, but he is paid to work, not to make important decisions. The captain decides when, where, and how to fish; he also is responsible for the safety of the boat and the deckhand.

A deckhand checks the condition of the nets after the last drag. The captain is visible to the right, pulling some of the gear over the side.

The captain may listen to the deckhand's opinion and advice, but he has the final say about the running, maintenance, and repair of the boat. Otherwise, the deckhand is most often the captain's coworker on the water, not his underling, even though he may be younger and less skilled. The nature and conditions of the work on a bay boat draw the men together; the work requires them to help each other out, to solve small problems through teamwork. If the two men do not get along well, the tension drives the deckhand to another boat, often with few words spoken between them.

Adding to this equality in the workplace is the probability that the deckhand is often related to the captain; he may be a son, nephew, or cousin. Sometimes he is the son of a friend or another member of the captain's fishing community. The ties and obligations of a captain to a family member or friend are restrictive and defined; if he plays Captain Bligh on the water, he knows full well that when the boat docks he will be held accountable by his family, friends, and other members of the community for his actions.

Fishing trips last from a few hours to several days, and the work is solitary, the captain and the deckhand forming a physically isolated team for the duration. The only possible social contact is through the marine CB. When bay shrimpers leave their home bay to fish, they are treated as outsiders by other shrimpers. Such trips bring the captain and deckhand more closely together, fostering a sense of comradeship based on mutual dependence.

The relationship between the captain and his deckhand also defines the quality of the work itself. Ultimately on a bay shrimp boat where the men are bickering, the productivity suffers. I hear frequently from captains that "a bad deckhand is better than no deckhand at all," an observation that emphasizes both the important work deckhands accomplish and the social role they play on bay boats. Larry figures that not having a deckhand on the *Queen Bea* increases the risk associated with fishing and probably lowers the amount of catch, but he increases his profits by not having to pay one. He tells me that he asks me to go fishing with him because he enjoys my company. It gets lonely out on the bay.

❖ ❖ ❖

SAFETY IN THE WORK PLACE

Texas bay shrimp fishing is not dangerous work, especially in comparison to other kinds of commercial fishing.[2] In answer to the question in the sample, "Are injuries common in bay shrimping?" the majority of respondents, 88.1 percent, said that in their opinion injuries are unusual. Many of the men view their work as dangerous only if certain commonsense procedures are ignored. A minority, 11.8 percent, disagreed, pointing to the ever-present problem of storms and heavy seas as a major threat to their safety. They are among the shrimpers who fish not only in the bays but also off the barrier islands in the Gulf of Mexico and are much more aware of dangerous weather. Both groups agree that a bay captain who fishes alone, without a deckhand, must be especially careful because there is no one to watch out for him if he falls overboard or if he gets hurt. Most shrimpers say that they often keep an eye out for each other's boats while fishing. Several men told of bay boat captains who fell asleep at the wheel while dragging their net; other fishermen on nearby boats would blow their horns and shine bright lights into the pilothouse until the captain woke up. Fatigue, especially at night, is a common problem; combined with the repetition and boredom of much of the work of fishing, it leads to mistakes in judgment. Hot coffee, caffeine, pills, and the companionship of deckhands help; moreover, a deckhand can allow the captain an hour or two of sleep.

The shrimpers were also asked if they had sustained any serious injuries while shrimping. (A serious injury, for the purposes of this survey, is defined as any injury requiring professional medical attention.) Only one out of 154 men in the sample said they had been seriously injured; he had fallen on his back while fishing and had spent three months recuperating before returning to his work.

Although according to these men serious injuries are rare, minor cuts, bruises, punctures, stings, muscle sprains, and abrasions are quite common. Shrimpers report that the most frequent kinds of minor injuries are stings from rays and puncture wounds from stepping on the spines of catfish. Working with the winches results in some minor injuries as do falls on the work deck. Falls cause minor back pain, but the fishermen claim that it almost always disappears after a week or two. Fish poisoning and rope burns are also mentioned as minor complaints. One fishermen reported that he broke his arm and another that he hurt his head while fishing the bay. The men usually doctor themselves; if an injury does not improve after a certain length of time, then they reluctantly seek medical attention. As one shrimper put it, "It's either going to get better or worse. And there's not a hell of a lot I can do about it. Besides commonsense. So you just wait and see what happens."

The self-reported-injury data seem reliable, but bay fishermen's recollections of the injuries of other fishermen must be viewed more cautiously. Two shrimpers reported seeing or hearing about two drownings, men who fell overboard from their vessels; because of the way data are collected it is possible that the two fishermen are recalling the same drowning. Six respondents know men who lost their fingers while working; again, it is unclear if the men are describing the same individual or different ones.

Only 1 percent of the sample has health insurance, far below the Texas state average of about 80 percent. The majority of fishermen say they want health insurance, especially for their families, but like many working Americans, they simply cannot afford it. A handful of fishermen say they do not require health insurance because bay fishing is safe work.

❖ ❖ ❖

WORK TENURE

Larry tells me with pride that he has worked as a bay fisherman since he got out of the army, about ten years. He is by no means exceptional; the men who work as captains have extensive experience in bay fishing and in other kinds of commercial fishing. The majority of bay shrimp captains in the survey worked in other Texas or

Gulf Coast fisheries before becoming full-time bay shrimpers. They had fished for oysters, finfish, and shrimp in the Gulf of Mexico before turning to full-time work in the bay. On the average, they worked in commercial fisheries for about 5.5 years before turning to bay shrimping.

The average fisherman in the survey has 13.3 years of experience netting shrimp in Texas bays, but the figure misrepresents the longevity of their work because it includes fishermen who are relatively new to the fishery, thus substantially lowering the total average years among those with much more experience. A frequency distribution provides a more accurate representation of the years of fishing experience of this labor force: 41.1 percent of the men have more than twenty years of experience as fishermen, 19.6 percent have thirty years, and 7.8 percent forty years or more. One of the bay captains has worked in the bay industry for more than fifty years.

The work-tenure data from the sample argue against a prevalent misconception the public holds that bay shrimpers are a come-and-go work force, that the majority of the fishermen demonstrate little permanent commitment to their work. The data in fact clearly show that bay fishermen are quite experienced in fishing. A corollary of this popular misconception is that bay shrimping is easy work, that anyone can become a bay shrimper. Such a belief trivializes the work of bay shrimpers, and people who maintain that view fail to understand the nature of the work on the water and the fishermen who accomplish it. The idea that bay shrimping is easy work is often predicated on the observation that entry into the fishery is relatively simple when compared to joining the offshore shrimp fishery. But the observation ignores the fact that entry into the bay shrimp fishery is only the first step in becoming a full-time bay shrimper. The people who venture into the industry soon face a hard reality, and experienced bay shrimpers watch with amusement as occasional newcomers give it up, usually after the first year. Bay shrimpers are puzzled when people who have never fished for a living think the work and the life easy.

Successful bay captains are skilled in all aspects of fishing on the bay; frequently they have fished other species and have many years of work experience. Unlike the newcomers who most often fail, they develop fishing strategies that are adaptive to the fishery and to the changing policies regulating it. Bay shrimping is not a part-time hobby, work that anyone can do; it is difficult and demanding work, even in the best of times.

There is no retirement age for bay shrimpers; they stop fishing when they decide they can no longer do the work to their own satisfaction. They expect to spend their working lives on the water, and when they are too old to lift the nets over the side or handle the winches or pilot their boats, they know it is time to retire.

❖ ❖ ❖

JOB SATISFACTION

Bay shrimpers' attitudes toward their work are diverse, complex, and entrenched.[3] Their views reflect their long commitment to shrimp fishing and their interpretation of the current problems confronting the industry. Bay shrimpers not only face a series of new governmental regulations but also a number of other problems that threaten to change the way they work; shrimp are harder to find and prices at the dock are falling. A majority of bay fishermen believe that the new laws, along with their economic problems, are destroying the industry. Whether they are satisfied with their work or not, they foresee a bleak future; according to one fisherman, bay shrimping "is a dying occupation."

One might reasonably expect a strong negative response to questions about work satisfaction, but such was not the case. Even when times are hard, the majority of bay shrimpers are positive about their work as commercial fishermen. In response to the question, "How much would you say you like working as a bay shrimper?" 84 percent of the sample said "very much" or "a good deal"; only 4.5 percent of the fishermen responded, "not much" or "not at all." To the question "Would you like your son (if you had one) to work as a bay shrimper?" 71.9 percent answered in the affirmative. And when they were asked, "If you had to do it all over again, would you shrimp for a living?" 71.4 percent said yes.

Open-ended questions on the survey invited some of the complexity and diversity of the responses. To the question "What do you like best about your work?" one shrimper answered, "You're independent. You can stay at it as long as you want. You can have a pretty good living." Another responded, "It's an outdoor activity. You can do what you want. You aren't on a time limit." Yet another said, "I'm my own boss. I work myself harder versus other jobs." And one shrimper said simply, "I love the water. Something different every day."

The dissatisfied minority answered the survey in a variety of ways. One man said, "I don't know how to do anything else. That's the only way I can make a living. It's too damn much work." When asked about the reputation of shrimpers locally, he answered, "It's bad. Don't ask me. I don't bother nobody."

The minority repeatedly mentioned the low reputation they feel they have in their community. One said, "They think we're killing all the shrimp in the bay," and another, "According to the game warden, we're all assholes." One respondent said, "The biggest number don't think much of us. Most think we are outlaws." And one shrimper commented, "They treat us like oil field trash."

❖ ❖ ❖

LAWSUITS

Lawsuits directly related to injuries from fishing for bay shrimp are rare. Although living in an increasingly litigious society, bay shrimpers do not rely on lawyers to solve their problems. When the survey sample was asked, "Have you personally ever sued because of an injury sustained on a bay shrimp boat?" not one individual answered in the affirmative.

Lawsuits among bay shrimpers are rare for several reasons. First, captains of bay boats frequently own the boats themselves; if they are seriously injured, which the data suggest is uncommon, then they have no one to sue. The boat manufacturer may be sued in certain situations, but often the captain builds his own boat or buys it from another shrimper; in the latter case, it is unlikely that the builder has sufficient assets to warrant a lawsuit. Manufacturers of equipment used on the fishing boats can be targets for a lawsuit, but bay shrimpers report that injuries from winches, which are particularly dangerous on Gulf shrimp boats, are minor and infrequent.

Bay shrimp boats are not usually insured. One might think that the larger and more expensive boats, those exceeding fifty feet, are more likely to be covered by policies, but even these are seldom insured. Bay shrimpers believe that the probability of their boats sinking or receiving major damage is low, regardless of size. The cost of insuring the boats is not seen as a good investment. And since owners of bay boats seldom carry boat insurance they are not promising targets for lawsuits.

If injured, a deckhand might be expected to sue the captain, but that does not often happen. Because the captain and the deckhand are sometimes related to each other or because they are friends, such lawsuits are rare. Family and community definitions of appropriate behavior clearly argue against a deckhand suing his captain.

Bay shrimpers rely on alternatives to lawsuits. If a captain injures himself while shrimping, he pays his own bills. If the boat is not owned by the captain, then the owner is expected to pay all or a significant part of the medical bills. A deckhand who is injured receives free medical care from the boat owner; often the captain, he considers it his personal responsibility to pay for any injury sustained by his deckhand. Along with financial support, the captain provides emotional support for the deckhand and his family. It is forthcoming not because of an implied legal obligation or the threat of being sued but because of the commitment of the bay captain to "do the right thing." After the fisherman recovers from an injury, his job is waiting for him on the shrimp boat. A different set of circumstances and outcomes prevails in the Texas Gulf shrimp industry, where legal suits are common.[4]

❖ ❖ ❖

BACK TO PORT

Larry and I head back to Port Lavaca on the *Queen Bea* around 8:30 A.M.; we had fished on the bay for more than twenty-four hours. By Larry's count, and I know he is always conservative in his estimate, we have at least 600 pounds of shrimp in the freezer, 200 pounds over the legal limit. The Alcoa plant has shrunk back to size, fading into the shoreline even as the port itself begins to grow in slow, even increments. The gulls return to haunt us; I no longer mind their pushiness and familiar loud cries.

There is a sudden noise to my left as I stand by the starboard railing and I jump back, reflexively; it sounds like a garden hose spurting when it is first turned on, air bubbles mixed with water. I cautiously look down; three feet below I see the exposed gray head of a dolphin, its mouth in a perpetual grin, top and bottom teeth showing. I hear Larry laughing behind me. The dolphin dives under the waves and resurfaces at the boat's stern, playing in the *Queen Bea*'s wake. Then I realize it is not the same dolphin, as out of nowhere there are ten dolphin fins all around us; just as quickly they disappear.

"Jesus," I say.

"They're something, aren't they," Larry says behind me, a statement, not a question.

The gulls grow louder as we draw into the small harbor. On the way in we pass four other bay boats headed out to shrimp, Larry waving to each one. As each one passes by Larry tells me the history of the particular boat and what he thinks of the captain's skills as a fisherman. We tie up at the same dock we'd disembarked from the day before. Larry jumps onto the dock and disappears into the fish house. I know he is going to check the price, to make sure that overnight it has not fallen substantially. He is in a good mood when he returns, talking with the dockhands, helping them unload the shrimp. The price is up two cents a pound.

"What time?" I ask Larry after I finish hosing down the boat.

"Be back here at 5:30."

I get my few things and climb onto the dock. I am hungry, tired, and looking forward to the long, hot shower back at the motel.

3

The Political Economics of Bay Shrimping

❖   ❖   ❖   ❖   ❖   ❖   ❖   ❖   ❖   ❖   ❖   ❖   ❖   ❖

The first thing Larry wants to find out when he returns from a night of fishing is the rate per pound for shrimp. The price fluctuates from one day to the next, enough to make a difference. He nets as many shrimp as possible to maximize his profit, but he knows that the price per pound, over which he has no control, determines his paycheck.

Larry and other Texas bay shrimpers do not earn guaranteed salaries; they are not paid for their time and effort or for the economic investment they have in their boat and fishing gear. Shrimpers total their gross, subtract their overhead, and figure their net profit while boarding their catch and culling through it. More than a few shrimpers tell stories of returning with large catches after a night on the water, only to find that the price has fallen and their hard work goes unrewarded. They believe that some of the more unscrupulous fish house owners listen to the shrimpers on their marine CBs, and if they hear the fishermen are catching more shrimp than usual, they lower their dockside prices.

Texas bay shrimpers are concerned about the regulations and laws that directly affect how and when they fish and, most important, how much shrimp they are allowed to net. They talk constantly about "the way it used to be in the good old days" and the way it is now. The shrimpers always keep an eye out for the game warden from the Texas Parks and Wildlife Department (TPWD).

Bay shrimpers believe there are two factors that determine their survival as commercial fishermen: the price they receive for the shrimp at the dock and the rules governing how to fish in Texas waters. They directly experience how price affects their wages from one day to the next and, at the same time, see and understand how these laws constrain them in their fishing for shrimp. These two factors, price and fishing laws, frame their economic world as commercial fishermen. Fishermen cannot influence or second-guess the biological processes that cause shrimp to be plentiful one year, rare the next; nature determines the number of shrimp in the bays. Nor can they necessarily understand the larger economic, social, and historical forces that may be at work. But they confront dockside prices and fishing regulations each fishing day, factors that are subject to their individual influence and that can be changed.

❖ ❖ ❖

THE SHARE OF THE CATCH

Bay shrimpers' annual incomes vary widely and are based on their share of the catch. After the captain receives the check for his catch (occasionally referred to as the boat's share) from the fish house owner, he figures out the amount to pay his deckhand and the amount he keeps for himself. In the survey most captains who use deckhands (92.2 percent) usually pay them 20 percent of the total sale of the catch at the dock; bay shrimpers sometimes refer to this cut as the crew share. If the captain is not the owner, then the sale of the catch is first divided between the owner and the captain. The most common arrangement, found among three out of four captains and their boat owners, is a fifty/fifty split of the gross sale of the catch.[1] From the captain-owner's or the owner's share of the catch are subtracted all costs for the maintenance, repair, and operation of the boat during and between fishing seasons, including the replacement of fishing gear, gas, ice, and personal supplies such as boots, gloves, and food.

A third of the captains in the survey who own their own boats also own another boat that regularly fishes the Texas bays. There are no owners who own small fleets of bay shrimp boats, however, as is the case with Gulf trawlers.[2] When the shrimpers were asked why a captain owns no more than two boats, one captain replied, "They're too many things you got to do with a boat. Boats take a lot of your time up. If you were going to run three, why you'd have to spend all your time just managing them. You wouldn't have time to fish." Many boat owners share the belief that two boats are the limit a fisherman can handle. The men who own more than one boat earn income from their own work as captain as well as the owner's share from both boats. An owner-captain usually considers his second boat a successful business proposition if he has a good captain running it.

❖ ❖ ❖

ANNUAL INCOME

People unfamiliar with the bay industry often believe that bay shrimpers get rich each year from their work on the water, but annual income figures from survey respondents disprove this common misconception. A certain segment of the labor force earns a comfortable living from fishing the bays, but the majority earn only a poverty-level income from it or are barely getting by.

The average earnings of a Texas bay shrimper in 1988 and 1989 were $19,859. A significant number of bay shrimpers in the survey sample (about 40 percent) earned annual incomes from shrimping that were below or just slightly above the poverty line. Another 40 percent were just getting by, their incomes from shrimping ranging from $15,000 to $29,000. At the other end of the scale, about 12 percent earned $40,000 or more. The diversity of income reflects the shrimpers' fishing skills and expertise, but it also mirrors the individual economic strategies they develop.

In order to supplement their incomes from bay shrimping, about one-half of the men (57 percent) net other species of fish during the year. Thus in addition to their incomes from shrimping, they can earn up to $9,000 more. Oystering is the most common source of income among these shrimpers; they fish for oysters not only during the short seasons that are permitted in Texas and Louisiana but also during times that are illegal. Several respondents commented that the oyster beds in Texas and Louisiana are rapidly disappearing, and they foresee their work and wages from oyster fishing to be fast coming to an end.

Oystering in both Texas and Louisiana is strictly monitored; licensing and fishing seasons are limited, fishing techniques restricted, and the allowable poundage per trip vigorously enforced. Nevertheless, many fishermen are apparently taking oysters illegally, according to the survey; they develop strategies to collect and sell oysters that circumvent the law and its enforcers. Texas bay shrimpers, along with other commercial fisherman, fish for oysters out of season and in beds restricted from commercial fishing; fish house operators do not ask where the oysters come from. Moreover, Texas bay shrimpers are not the only fishermen to participate in illegal fishing.[3]

About one-quarter of the bay shrimpers (24.3 percent) earn income from work that has nothing to do with commercial fishing. They work "off-boat" jobs and earn a gross average of $8,077 annually to supplement their income from commercial shrimping. Bay shrimpers who take off-boat jobs typically labor at blue-collar work that pays only a few dollars above minimum wage. They work these jobs between fishing seasons or when the shrimping is so bad that it makes no economic sense to continue fishing, each fisherman deciding for himself when to stop and to tie up his boat at the dock. The off-boat jobs include welding, clerking in convenience stores, and a variety of other jobs that the fishermen view as necessary to supplement shrimp fishing income in good and bad years, as something to fall back on.

Only three bay fishermen in the sample hold off-boat white-col-

TABLE 3.1

Gross Annual Income of Bay Shrimpers, 1988, 1989

Gross Annual Income ($)	Bay Shrimping (N = 120)	All Sources (N = 123)
2,000–4,999	4.1%	2.4%
5,000–9,999	20.0	11.3
10,000–14,999	15.8	13.8
15,000–19,999	17.5	15.4
20,000–29,999	20.0	20.3
30,000–39,999	10.8	16.2
40,000 or more	11.6	20.3

lar jobs, in each case a lower management position with a nearby petrochemical company. The three men fish the bays on weekends, during vacations, or sometimes after a full day's work. After retiring from their white-collar jobs, they each plan to spend longer hours fishing in the bay. Yet they do not intend to rely on their fishing incomes in order to live; they count on their retirement income from their company jobs. They view bay shrimping as more than just a weekend's recreation but as much less than a full-time occupation.

There is a strong belief among some coastal residents and policymakers that bay shrimping is part-time work. The part-timers, according to the myth, work at their "real jobs" most of the week, then sporadically fish the bays for shrimp when time allows, the extra income supplementing their regular salaries. Although the data do not support the notion, it is quite commonly held, serving the function of trivializing the work and the lives of the bay shrimpers. The misconception contributes to the stereotype that bay shrimpers are by definition of little consequence.

When all other sources of income were considered, the mean average gross income of bay shrimpers in 1988 and 1989 was $25,443 (see Table 3.1). The range of their gross total incomes was even more diverse than the range of incomes derived from bay shrimping alone; in the sample one individual earned $3,500, another $85,000. When all sources of income are considered, about 9 percent more shrimpers appear in the upper-income category. At the same time, the percentage of shrimpers at or near the poverty level declines from about 57 percent to about 43 percent when all sources of income are included. Their annual earnings reveal why the Texas bay shrimp fleet looks like such a rag-tag affair, with so many boats in need of repairs and maintenance. Many bay shrimpers do not have the money to keep their boats shipshape; only about half the shrimpers in the sample made enough money to maintain their boats, to make major repairs, and to purchase up-to-date navigational gear and other electronics.

Coastal residents often complain about the large government subsidies that bay shrimpers receive when they cannot fish the bays, but the data do not support the frequent criticism. Although a large percentage of bay shrimpers qualify for some form of government assistance, few choose to receive it. No shrimpers receive Aid to Families with Dependent Children (AFDC) or Aid to the Blind and the Disabled (ABD), and food stamps and Supplementary Security Income (SSI) are used by only 2 percent of the sample. Only one shrimper out of 150 reported that he receives unemployment insurance.

Bay shrimpers try to get by during bad economic times; those with earnings from jobs in addition to shrimping clearly are in a better position. Thus the shrimpers look for other kinds of work when necessary, depend on family savings, and turn to close family members and friends for financial help. Always hoping for better times, the men actively seek ways to weather their economic storms instead of relying on what they call "free handouts."

❖ ❖ ❖

THE ECONOMIC IMPACT OF BAY SHRIMPERS

The total gross annual incomes of Texas bay shrimpers have a significant economic impact on their communities. The 120 shrimpers in the survey on whom data were collected earned about $2.4 million; earnings for all Texas bay shrimpers were estimated at $138.7 million. Moreover, the work of bay shrimpers supports additional jobs and incomes, including dockside workers, fish house operators and workers, and truckers, which in turn generate additional dollars in the local economies. Using a suggested income multiplier of 2.37, the labor force of bay shrimpers has an economic impact of $328.7 million.[4]

The much larger Texas Gulf boats catch far greater numbers of shrimp than the bay boats. Nevertheless, the data suggest that bay shrimpers earn somewhat more from their work on the water than do their Gulf counterparts. The differences in wages derive directly from the wage agreements of Gulf and bay shrimpers, the ways in which the sale of the catch is divided.

Bay deckhands are the equivalent of Gulf headers, but Gulf rigmen have no counterparts on a bay boat. Captains on Gulf and bay boats bear the same responsibilities for their boats and crews and make similar kinds of decisions. On Gulf boats, however, the crew wages are divided according to a system that follows the traditions of Florida and other Gulf state shrimpers; Texas bay shrimpers have a different set of work traditions. Generally, on Texas Gulf boats headers are paid a piece-rate; they receive an agreed-upon fixed dollar amount per hundredweight, or box, of shrimp that they head or help to head. The best of the headers sign on with the captain and are paid twenty-five dollars or more per box of shrimp tails. Rigmen, however, earn a percentage of the boat's share of the gross sales of the shrimp at the dock. The better the rigman, the higher the percentage he demands and receives from the captain. The best Gulf rigmen split the boat's share with the captain forty-five/fifty-five, less the wages of the headers, minor boat repairs, and possible percentages of the cost of gas and ice.

A Gulf captain, who rarely owns his own boat, makes a verbal agreement with the boat owner before he ever leaves the dock. Most commonly the captain's earnings are a 35 to 65 percent cut, the majority share going to the boat owner; the captain's, or boat's, share is

TABLE 3.2
Annual Income of Bay and Gulf Shrimpers, 1988–89

Annual Income ($)	Bay Shrimpers (N = 120)		Gulf Shrimpers (N = 89)		
	Deckhand	Captain	Header	Rigman	Captain
2,000–14,999	89%	36%	100%	79%	30%
15,000–29,999	0	41	0	21	65
30,000–39,999	11	11	0	0	4
40,000 or more	0	13	0	0	0

then divided between the captain and his rigman, less costs.[5] Estimated at less than 10 percent of all Gulf captains, the few men who own and operate their own boats are among the highest income earners in the Texas Gulf fleet. Gulf shrimpers rarely take off-boat jobs.[6] There are other ways to determine wages on Texas Gulf boats, but they are far less common.[7]

In spite of the Gulf shrimpers' higher annual production, bay shrimpers are more likely to earn in the higher income categories.[8] Almost a quarter of all bay captains (24 percent) earn $30,000 or more compared to 4 percent of the Gulf captains (see Table 3.2). And bay deckhands earn more than their Gulf counterparts; however, the small number of deckhands in the sample requires caution in making the comparison. The differences in incomes are the direct result of the wage-share agreements. In short, the owners of Gulf boats receive a greater percentage, and thus a larger dollar amount of the gross sale of the shrimp at the dock.

Boat ownership is the key, and bay shrimp captains most often own their boats. Maintenance, repair, gas, ice, and other costs are relatively small on bay boats, compared to the overhead on Gulf boats; still, the costs are borne by the captain-owner. Gulf crews often pay a percentage of the costs for equipment repair, gasoline, ice, and food, which takes a hefty chunk from their checks.

Bay captains who own and fish the boats and bear all economic risk are small-scale entrepreneurs. In contrast, their deckhands, and the entire fishing crew on Gulf boats, including the captain, are wage earners, members of the labor force of commercial fishermen. Unlike the Gulf boat owners who give over responsibility for the boat, the equipment, and the crew to their captains as soon as the bow lines are cast form the dock, the majority of bay captains maintain complete control over the fishing effort from start to finish. Bay captains determine their own fishing schemes, strategies that can maximize their profits.

❖ ❖ ❖

FISHING STRATEGIES

The economic and political world of bay fishermen is complex, more intricate than is indicated by their notion of price and fishing regulations. Although the work that bay shrimpers engage in on the water is technologically unsophisticated and requires minimal formal education, these small-scale entrepreneurs nevertheless are enmeshed in regional, national, and global market forces. Moreover, continuing changes in the regulation and enforcement of the bay industry, often driven by forces outside the purview of the fishermen, ensnare them in a web of conflicts that requires careful resolutions.

Bay fishermen do not escape the complexity when they dock their boats. Shrimpers make choices on the water that affect their ability to profit from fishing, but they also make choices within their households that directly influence their work. Shrimping permeates every aspect of their lives. Choices range from the selection of certain kinds of fishing gear to maximize the amount of the catch to the less obvious ones involving what kind of work the fisherman's wife might do, child-care arrangements, and the allocation of household tasks.

The choices that fishermen and their families make, their decisions in response to the economic and political complexities they face because shrimping is their main livelihood, are rational and form clear-cut strategies as their annual incomes reveal. Strategies are of three different sorts. The first includes methods for shrimping and fishing that more or less follow the laws and policies set forth by the governing agencies monitoring the commercial shrimping industry. The second strategy is to maximize annual earnings from shrimping through illegal methods intentionally structured by shrimpers to circumvent the prevailing laws and policies dictating bay fishing. The third strategy is to take off-boat jobs that allow them to maintain shrimping as a way of life. Most of the off-boat jobs are legal; a few are not.

Legal Fishing Strategies

Bay shrimpers are highly regulated in their commercial fishing in the Texas bays.[9] They are told which seasons and times of day and night they are allowed to fish, how much shrimp they may net, what types of fishing gear and equipment are permissible, and where they may fish; and they are mandated to uphold certain safety regulations. Throughout their work on the water, they are told what is permitted by law and what is not; when they return to the docks from fishing, their boats, equipment, and catch are subject to inspection.[10] The state and federal laws are enforced by agents of the Texas Parks and Wildlife Department and by the U.S. Coast Guard. The state regulates the bays and river systems and the Gulf waters to three miles off the barrier islands, and the federal government controls fishing waters from three to two hundred miles off the Texas coast.

Bay shrimpers can purchase four kinds of commercial shrimp fishing licenses. They are required to purchase a general commercial fisherman's license, which costs $15 for state residents, $100 for out-of-state. Then they can also purchase a bay shrimp license priced at $80 for state residents and $320 for out-of-state; it allows a captain to fish his boat during two seasons each year, spring, which begins May 15 and ends July 15, and fall, from August 15 to December 15. Bay shrimpers may begin shrimping the major bay waters of Texas, the "inside" waters, thirty minutes before sunrise and must pull up their nets thirty minutes after sunset. This second license restricts them to day fishing only. A catch limit of no more than 300 pounds is permitted. Shrimp are most abundant during the spring season, migrating slowly from the bays into Gulf waters. During the fall season the shrimp are usually far less abundant, except in the beginning weeks of the season.

A bay shrimp license also permits commercial fishermen to fish in the waters of the Gulf of Mexico to three miles offshore. These "outside waters" are open to bay shrimpers all year except June 1 through July 15. When shrimp from the bay are docked, they are sold to the fish house operators who hold shrimp house operator licenses, purchased for $300; the operators may purchase shrimp only from commercial fishermen holding a bay shrimp license. Shrimpers' fishing gear and equipment is restricted, including the mesh size of the try net and the main fishing net. The size of try nets, otter trawls, doors, and other important gear is precisely determined and regulated as is the size of the shrimp netted. From August 15 through October 31 bay shrimpers are allowed to take shrimp that are no smaller than fifty per pound; shrimp are counted before they are deheaded. During other times of the year there are no size restrictions on shrimp.

The third kind of commercial fishing license is a bait shrimp license, which permits a shrimper legally to net up to 200 pounds of shrimp at any time of the day or night for most of the year; the one exception is during the fall season from sunset to sunrise. At least 100 of the 200 pounds allowed by a bait license have to be kept alive in the well of the boat although the rule does not apply from August 16 through November 15. The shrimp are docked and then must be sold to bait house owners or directly to sports fishermen. The former must be licensed as bait dealers in order to sell bait shrimp; they sell them to recreational fishermen who use them as lures for trout, redfish, and other coastal game fish. Some recreational fishermen prefer their bait to be alive, thus the rule that at least half the bait shrimper's catch must be kept alive. The other bait shrimp are frozen and then sold in one-pound or smaller containers to recreational fishermen. Live bait shrimp are kept in tanks at the bait stands until they begin to die; then they are frozen, placed in containers, and sold as bait.

Some bay shrimpers, a minority, also purchase licenses that allow them to fish from three miles out in deeper Gulf waters. A Gulf shrimp license, costing $100, is similar to the bay license in regard to seasonal and gear restrictions, but it does not have catch limits; Texas Gulf shrimpers may net as many shrimp as they can bring over the side of their boats.[11]

A major objective of the licensing laws and policies is to limit severely the number of shrimp that the shrimpers may net in Texas bays. For example, even in the spring season, when shrimp are most abundant, fishermen are not allowed to net more than 300 pounds. Legal and policy distinctions between bay and bait shrimpers and the shrimp they catch have little basis either in the work shrimpers accomplish on the water or in biological fact; the intent of the laws is to restrict the amount of catch. In fact, the fishermen use the same kinds of boats, gear, strategies, and fishing techniques; the only difference is that bay shrimpers must sell their shrimp to fish house operators and bait shrimpers to bait house operators. In practice both groups sell to either, depending on the time of year.

The bait shrimp business requires a steady flow of product during certain peak times of the year. Bait house operators, who are often bay shrimpers themselves, use their own boats to meet the need or hire other bay shrimpers to sell their catch to them on a regular basis when the demand from recreational fishermen is high. During slack times of the year, bay shrimpers sell to the fish house operators.

The separate bay and bait licenses also foster a cultural misunderstanding of shrimp. Based on their own experience, many coastal Texans know that bay and bait shrimp are the same and consider both kinds as fit for the dinner table. The coastal Texans know that if bait shrimp are not used up in fishing for trout, then they are brought home and put on the stove to boil, so that even if no fish are caught, fishermen can still have fresh shrimp for dinner. But Texans and others who rarely visit the coast believe or come to believe that shrimp designated as fish bait are different from those that appear in their shrimp cocktails. The confusion of the majority of the public helps sustain irrational fishing laws and policies over which bay shrimpers have little power or control.

Bay shrimpers do not accept the ways in which bay and bait licenses are framed by industry regulators or, most important, their intent. Many bay shrimpers are determined to avoid the fines and penalties imposed for breaking fishing laws and at the same time to dock the largest daily poundage of shrimp legally permissible. To this end, they buy licenses for bay and bait shrimping; a few also purchase a separate Gulf license. They then develop strategies to land a minimum of 500 pounds of shrimp legally during the spring season and during other times of the year to increase their catches.

A bay shrimper can net 500 pounds of shrimp legally during the spring season although the intent of the law is to limit him to a bay license (for 300 pounds) or a bait license (200 pounds). Once shrimpers have caught their limit of bay shrimp, they sell the 300 pounds to the fish house. At sunset the same day, the shrimpers motor out into the bay again, legally defined as bait shrimpers by their second license. By law the bait shrimpers pull in drags of bait shrimp. As required, they place half the 200-pound limit into a live well; the other half is preserved on ice. Once the shrimpers dock, they are no longer required to keep the bait shrimp alive, so they switch off the air pumps attached to the wells.

The captains and deckhands sort the shrimp into plastic baskets and carry them to the bait house dealers, who may be the same men as the fish house operators; shrimpers sell their catch at the same price per size as they received for the bay shrimp earlier in the day. The dealers then sell the bait shrimp to recreational fishermen or place them in freezer trucks in baskets or cartons, side by side with the shrimp docked earlier.

A small number of bay shrimpers with a bait license sell shrimp to bait houses during certain times of the year, but the majority who purchase a bait shrimp license do so only in order to raise their maximum legal catch from 300 to 500 pounds per twenty-four-hour day. They may not follow the spirit of the law, but they certainly follow the letter of it. They are not stopped, fined, or arrested for this practice. Everyone I spoke with, shrimpers, fish house operators, dockhands, fishery agents, and other people familiar with the bay industry, are well aware of the strategy; indeed it is prevalent among the bay fishermen in the twenty different communities in the sample. They believe it is necessary to earn a legitimate profit from their work.

When I asked fishermen if they could make it on a 300-pound limit during the spring season, I was often laughed at, cajoled, and then patiently shown the basic economics of bay fishing as they knew it, often figures jotted down on a stray piece of paper. Some fishermen wrote down the numbers, circling the important figures; others, after writing them down, just handed the paper fragments to me, the logic of their argument so obvious that to them there was no need for clarification or translation. Others recited by rote what they needed to earn on a regular basis in order to make it. Although a majority of the shrimpers felt that a 500-pound limit was enough, some argued that even their profits from 500 pounds were "a fool's wage." One fisherman said frankly, "I don't want to do it this way [using both bay and bait licenses]. It's like lying. But you tell me . . . what can I do?"

Other fishing strategies are used to maximize the catch. Bay fishermen work on the water netting shrimp a mean number of 6.5 days per week, and more than 60 percent of the respondents reported that they fish at least six days of the week during fishing seasons. The figures suggest that the shrimpers make every effort to net as many shrimp as they can by working long hours during the seasons, taking only an occasional day or half day off.

More than three-quarters of the bay shrimpers keep daily and weekly catch records, either by keeping the sales receipts from the fish house operator or by writing down a running total of their gross from their catch. Four out of ten shrimpers in the sample have accountants or bookkeepers who help them with the paperwork. As small, independent businessmen the fishermen know exactly how they are doing at any point in the year, how much they are in debt, if any, and how much they need to make in order to match the previous year's income. The data further disprove the myth that bay shrimpers are slipshod part-timers who fish only in order to supplement income from other full-time work.

Bay shrimpers use bay and bait licenses to fish as long as it is profitable for them as entrepreneurs. Bay shrimpers fish 7.23 months per year on the average although the maximum fishing time possible using a bay license alone is six months (bait shrimpers can fish all year round except at night during fall season). When the shrimp are plentiful, the men fish the bay waters every day that is legally allowed and as many as they can tolerate. If they need the money, they fish until they are physically exhausted or too tired to take their boats out safely or until their boat requires major repairs.

Some bay shrimpers also use their Gulf shrimping license, following the shrimp as they migrate from the bays into the Gulf of Mexico. Although the shrimpers spend most of their time fishing the bays, rivers, and inlets, 53.2 percent of the men in the sample also fish in Gulf waters. Traditionally, bay shrimpers who work in the Gulf of Mexico fish "off the beach," no more than three miles from the shores of the coastal barrier islands in state-controlled waters, and they are more likely to fish these waters during quiet days when the weather conditions are at their best, the winds light, the chop less than a foot. The shrimpers who do not fish the Gulf (46.7 percent of the sample) say that they are constrained from doing so by the capacity of their boats, which are not built to weather Gulf waters. The definition of their work as bay shrimpers also limits them to the bays; fishing in the Gulf of Mexico is not a part of their work even if shrimp are abundant off the beach because that is the territory of Gulf shrimpers.

Yet bay shrimpers feel increasing economic pressure that is causing them to change their definition of traditional fishing grounds. About one-fifth of the sample (19.3 percent) said that they fish offshore, not directly off the beach but in the deeper waters of the Gulf of Mexico. About 5 percent reported that they did so "often," 15.3 percent said "sometimes." They are expanding their territory in an attempt to maximize their catch. Outside state waters there is no mandatory limit on the amount of shrimp that can be netted. By crossing over into another distinct fishery, they catch and earn more than by fishing in the bays, rivers, and off the beach. They still define themselves as bay shrimpers, however. They also are broadening their fishing territory in another way. Traditionally, they have fished one bay system, their "home" bay; now some are migrating to other Texas bays to fish for shrimp.[12] The fishermen temporarily move their base of operations to a nearby bay system when their home bay stops providing a reasonable profit; when the shrimp return to their home bay, they move back.

Vietnamese bay shrimpers commonly practice this fishing strategy, and larger numbers of non-Vietnamese shrimpers apparently are doing so. The Vietnamese live on their boats when they fish other bays, but Texas bay shrimpers live in campers or small trailers, or they rent rooms close to the dock. Texas shrimpers do not like moving to another bay to shrimp, not only because they miss their families and friends but also because they usually do not know how to fish the foreign bay. Moreover, they often do not know the local bay shrimpers and cannot depend on them for useful information about the shrimping or for assistance if they need to repair their boats.

Texas bay shrimpers who do stay in their home bay must still find shrimp. David R. M. White, focusing on the strategies that shrimpers use to help them find the shrimp, studied Alabama shrimpers to understand some of the advantages and disadvantages of fishing alone and in groups of boats.[13] Texas bay shrimpers resolved the problem of where to fish in a number of ways, just as shrimpers in Alabama did. The Texans drew on a variety of information to decide where to fish, including their own experience, conversations with other fishermen before leaving the dock, and communication with fishermen on the CB while fishing. The resolution of the problem facing bay shrimpers is played out within the framework of their group dynamics, the nature of their relationships with other fishermen both on and off the water; the dynamics between and among fishermen are intricate, reflecting the full range of emotions that is tied to the importance of fishing for a living.

One group of bay fishermen I observed for seven consecutive days and nights of fishing is composed of sixteen to nineteen captains, not including their deckhands. Several of them are held in high regard by the other fishermen; they are much better producers than average and are often asked where they fish. Their advice is sometimes taken, sometimes studiously ignored. A bay captain such as Charlie, for example, does not base his decision to fish simply on what Joe, who others say catches more shrimp than most, may suggest. A number of possible factors may come into play in Charlie's fishing decision, including how he feels about Joe. For instance, if Charlie owes Joe money, or if Joe insults Charlie's sister-in-law at a party in town, or if Charlie's deckhand is Joe's second cousin, then Joe's expert advice is either proportionately credited or discredited, taken or ignored.

Within this group of fishermen are cliques bound together by ties of friendship, kinship, or both. Captains in cliques fish together, at least most of the time, but there are exceptions. One night, for instance, a captain might "get a hair up his ass" to fish by himself or with someone outside the clique. I asked one captain how he makes the decision on where to drag his nets: "Well, it depends. On a lot of things. Sometimes I follow Old George, then sometimes I don't. I go where he goes because he catches his fair share and then some. But then they'll be times he is just going on a hunch. And I don't think I should [follow it]. I might go out with Bill, he's one of my fishing buddies, or just go off by myself. It all depends." Another captain I fished with takes a different view of his fishing strategies: "I know where the shrimp are, if they are out there, and I go right to where they are. If I'm wrong, I'm wrong. Big deal. I lose a few dollars that night, but I'll make them up the next time. If the other guys in this bay want to fish with me, they know I'll do my best. They take their chances like anyone else. There are no guarantees in this fishing business." When I asked him why he does not follow one of the other groups of boats that fish together, he grew disdainful: "I'm not out here to have somebody tell me what to do. If I want that, I can work over there. [He motions at the large refinery two miles to the northeast.] I go where I think there's shrimp. That's about it."

Some shrimpers, then, follow the advice of some captains; others may intentionally ignore such advice on an irregular basis. And still others, if there is no reliable information on any particular day or night, may decide to fish "where they feel like it." I asked one captain, who says he fished in the bays for more than twenty-five years, how he decides where to fish. "It depends," he said. "Depends on what?" I asked. "On how I'm feeling, the weather, what I got to do that day, those kinds of things." "Have you ever flipped a coin to decide which way you're going to go?" I asked. "No, not a real quarter. But sure, we might just be joking around, not see where one place was any better than another. And I might say out of nowhere, let's go to such-and-such a place and give it a try. So we would do that. Now I don't do that much, but sure, you get to the place where, shoot, it probably don't make that much difference where you go. Like when there is not much out there and they're real small."

The majority of the captains in this group have strong opinions about the fishing expertise of their fellow fishermen. They truly believe that some captains are better than others, that for a variety of reasons they can net more shrimp every year than other men. The debate about the so-called "skipper effect" has raised the interest of several researchers.[14]

Another legal way by which shrimpers maximize their annual incomes is to build their own boats. Besides saving on their capital investment, they benefit from a more complete knowledge of the operation and maintenance of the boat, saving themselves money in the long run. They fish from boats they know by heart. Each boat has its own intricacies, and a knowledge of the boat directly leads to less down time for repairs and to higher production. In contrast, captains who buy their boats initially know the idiosyncrasies of the boat less well and of course must pay back the principal and interest. Although boats are inexpensive, especially when compared to the much larger Gulf boats, one or two bad years in a row can turn a small bank loan into an economic burden.[15]

One way to lower overhead is to save money on fishing gear and repairs to equipment. Some captains in the survey sample said that they spend a significant amount of time finding and purchasing the best, and cheapest, buys in equipment. They tell of driving their families in their pickups to the nearest metropolitan area, which may be as much as fifty to seventy-five miles away, to get a good deal on a piece of gear. One captains knows, for example, that a Walmart, twenty-five miles from his village, is running a sale on large coolers; he will save a few dollars per cooler over the price he would have to pay at his local marine-supply store. Another captains reports that he always goes "to the city" for his mechanical repairs or for new equipment because he believes the people there do better work and charge less. Other captains buy in quantity, whether marine paint or rubber gloves, buying items during sales from local merchants; word gets around quickly from one captain to the next if there is a special sale going on in town. Some captains do their own welding and much of the mechanical work, such as overhauling the winch motor and keeping the boat engine in running order. The younger captains in the sample are less likely to have these money-saving skills than are the older ones.

To lower overhead significantly, some captains get rid of the deckhand. During hard times the extra 20 percent saved from eliminating that position can be enough to make a big difference. Although not using a deckhand increases profits, it also can cause problems; only 12.5 percent of the sample reported that they do not use deckhands regularly to help them fish. Captains rely on a wide variety of methods to operate and fish the boat without a deckhand. As inventive as some of the methods are, from using a broom handle to serve as an automatic pilot to buying sophisticated navigational aids, bay shrimpers are going to catch more fish and will do so more safely when there is a deckhand working the boat with the captain.

A few bay shrimp captains in the sample, 5 out of 142, hired their wives as deckhands; the women offer an extra pair of hands useful when the catch is hauled over the side of the boat and during culling. Wives are also free labor, an important savings to the captain and his family of 20 percent of the gross sale of the catch.[16]

Illegal Fishing Strategies

Many of the shrimpers believe they cannot make a decent living as long as the current laws and policies are in effect to regulate fishing in the Texas bays. In their view, the regulations directly translate into fewer shrimp netted, fewer shrimp sold at the dock. They believe their only option is to break the existing laws, which they see as unrealistic and as the product of their political enemies. The men feel that circumstances force them to become lawbreakers, yet their rationalization does not seem to minimize their frustration and anger.

"Can you believe this?" one captain asked me as he went about his nightly fishing. "What I'm doing here, why . . . I'm a goddamn outlaw according to Parks and Wildlife. If that ain't a crock of shit, I don't know what is." Shrimpers who regularly break the fishing laws are not considered criminals by other shrimpers, their families, or many of the members of the community who are not fishermen. "You do what you gotta do," one fisherman told me when I asked him how he felt about the shrimpers breaking the laws.

The most common illegal fishing strategy is to catch more shrimp than allowed by the licenses. The shrimpers always try to maximize their catches; the more shrimp, the greater their profits. In practice, bay shrimpers frequently exceed the catch limits on both their bay and bait licenses. A fisherman may net far less than 300 pounds during the Bay for several weeks; then, when the shrimp are running, he may get his catch limit after only two drags. He continues fishing, taking his excess shrimp into the dock with one eye out for the game warden. Returning to fish on his bait license, he again may fill his nets when the shrimp are abundant, landing enough for two boxes early in the night. By law bay shrimpers must return to the dock with their catch limit, but on several occasions during the course of the research I observed shrimpers fishing for hours after they had exceeded their limits. Bay shrimpers told me in detail about their illegal fishing strategies. "How can I stop?" one fisherman asked. "It hasn't been this good for three weeks." He had brought to the dock over 700 pounds of shrimp in less than twenty-four hours, 200 pounds over the bay and bait limits.

When a shrimper catches more than the law allows, it is a cat-and-mouse game between him and the agents of the TPWD. Texas bays and rivers are large expanses with few if any agents monitoring a particular bay. Shrimpers communicate with each other by CB, often using simple code words, whenever they see a possible TPWD boat or agents trying to spot them with binoculars from land. It may be a cat-and-mouse game, but there are usually few cats. Bay fishermen use simple strategies on a trial-and-error basis to trick their pursuers, relying on whatever works. After exceeding the catch limit, shrimpers get on the CB to see if their friends have spotted any agents in the vicinity; they also check the radar if they have one to locate any suspicious boats. Then they make a dash to port, hoping for the best; of course, it is in reality a slow and cumbersome return to port in boats that top out at ten knots or less. Shrimpers keep their eyes on the horizon for marine agent boats and any signs out of the ordinary. If luck is with them, they make it safely to the dock where they offload as quickly as possible. Agents are easy to identify, dressed in their official uniforms, driving identical trucks with state agency insignias on the door panels.

Bay shrimpers have individual ways of responding to the sudden appearance of the TPWD agents. If a shrimper sees the agents from a distance, he usually dumps the excess shrimp over the side of his boat away from the approaching boat. The men hate to do this; one complained "The shrimp are already dead so what difference does it make?" After dragging for hours, it angers them to have to "waste" the shrimp after all their effort. In a few seconds' time they get rid of any evidence of their wrongdoing; the agents board their boats, check their licenses, estimate the amount of shrimp, then leave. If shrimpers do not have time to dump their shrimp, they try to "talk their way out of it," pleading with the agents to "give us a break, let us do our work." The strategy frequently does not work, but the men often attempt it as a last resort.

If all else fails, shrimpers simply admit their guilt and even help to hand over the shrimp to the agents. They get a ticket, show up in court when required, plead guilty, and pay the fine. They then try their best to make up the costs of the fine and their confiscated catch the next night by again breaking the law. Once they recoup their losses, they are especially careful not to get caught again. If shrimpers are stopped at the dock with the extra shrimp, they have fewer options than if stopped in the bay; one shrimper said, "I just roll over and play dead. They got me."

Another common illegal fishing strategy is referred to as "mixing shrimp." Bay shrimpers are not supposed to net or sell undersized shrimp, those weighing less than fifty per pound, from August 15 through October 31; they are juveniles that will mature in size, gaining in value if netted later. Yet because of the new laws, bay shrimpers often are in a situation in which it makes more economic sense to catch some shrimp, even if undersized, than none at all. If given their choice, they certainly prefer to net legal shrimp; but if the undersized are the only ones to be netted, then it is better to catch illegal shrimp and keep their boats operating, their gear and equipment in repair.

During certain times of the year if bay captains cannot find larger shrimp when they test the waters with their try nets, then they net the undersized ones even as they continue their search for legal shrimp. If luck is with them, they eventually net some larger shrimp, which are then mixed with the smaller ones. If the shrimpers are stopped and boarded by the TPWD agents, the fishermen show them the mixed shrimp. The average of the mix must fall below the minimum size requirements for the catch to be illegal. The fishermen contend that the shrimp were netted at the same time so that no law was broken; the agents know the trick but can do nothing about it.

If the fishermen are less lucky and do not bring aboard larger shrimp to mix with the smaller ones, then they run the risk of being fined for netting shrimp that are undersized. To avoid fines, they wait for the right time, then hurry back to port and quickly offload the shrimp. The fish house operator buys the undersized shrimp, knowing it is illegal and mixes it with larger shrimp he has already purchased. If the TPWD agent approaches the shrimper on the bay with a load of undersized shrimp, he will dump the fish over the side to get rid of the evidence.

Agents sometimes check the records of the fish house operator to ensure that bay fishermen do not sell more than 500 pounds of shrimp every twenty-four hours. Shrimpers, however, have developed a variety of ways to misrepresent their actual catch on boathouse records to avoid fines. One common practice is to offload shrimp at more than one boathouse. For example, should a bay shrimper catch 800 pounds of shrimp one night, he may sell 200 pounds at one fish house, 300 hundred at another, the remainder at still another. If the port is small and his actions too obvious, then he may offload some of his shrimp at the dock and take the rest to another port to sell.

I also observed bay shrimpers giving their catch to friends who then sell it under their name to the fish house operator. Another strategy is to sell only 500 pounds to the operator; then the fishermen return the next day but leave their boats idle at the dock and sell the remainder from the previous night's catch as shrimp they have netted that day. Yet another option is to sell the shrimp themselves; however, the practice is unpopular since it takes away from the shrimpers' time on the water. In short, there are a variety of illegal ways to get around the fish house operator's recordkeeping and to evade the watchful eye of the marine agent.

Bay shrimpers occasionally use fishing gear that is prohibited by law. They fish with larger nets than are allowed or, more rarely, with nets whose mesh is smaller than required. Although they can catch more shrimp with illegal nets, they do not use them often because it is too easy to get caught. To catch shrimpers red-handed, agents measure the nets while they are still wet from use. It is impossible to change from an illegal to a legal fishing net quickly without being observed by an approaching TPWD boat.

Bay fishermen sometimes fish out of season or in waters that are closed to shrimp fishing, a strategy that might net hundreds of pounds of shrimp, but the real problem is how to sell the catch. During closed shrimp seasons, a fisherman has no reasonable explanation for how he acquired his catch, nor does the fish house operator who buys it from him.

Shrimp fishermen do not like to see their competitors fishing out of season or in closed waters. One bay shrimper told a story about one night when he and several of his friends, each with a boat, were fishing in a river that was closed to shrimping at the time. When they returned to the dock to sell their shrimp, they were surprised by TPWD agents who confiscated their catch and gave them each a ticket. The men were angry to find out that another shrimper had called a hotline number to report them. "What he did was wrong. There was no way he should have done that. [He] is pure crazy."

Shrimpers who report their fellow fishermen, and nonfishermen who keep bay shrimpers from pursuing illegal strategies, are automatically defined by the lawbreakers as the enemy. And TPWD agents, the Coast Guard, and county and city law enforcement officials are viewed alike by the fishermen as part of what is wrong with bay shrimping today.

Much of a long night's conversation on the CB among bay shrimpers, talk that might go on for several hours between drags, focuses on the locations of law enforcers and their recent activities. But there is usually little specific information available, always much conjecture and rumor. Rarely is there consensus on what actions if any can be taken. Often a fourth, fifth, or sixth shrimper will add his comments to the ongoing conversation, taking the freewheeling talk in a totally new direction.

Constant, if haphazard, vigilance and cooperation prevail

among the fishermen, who share a common brotherhood in their work and style of life. Some shrimpers are ostracized, however; the shrimper who was described as crazy by his peers is unwilling to see other fishermen net shrimp out of season. Most shrimpers help each other any way they can, especially against a perceived common enemy, the Coast Guard; at the same time, one fisherman's gain can be seen as another's loss. The men hate agents of the TPWD and those who serve in the Coast Guard, but they also at times detest and distrust each other.

I observed one illegal fishing strategy that requires the cooperation of more than ten fishermen, a fish house owner, and his dockside helpers. Several nights each week of the summer fishing season these bay shrimpers play out the same act. First, they dock their boats by about 5:00 P.M., each captain making sure, if other people are around, that the nonshrimpers understand that the shrimping is terrible, not worth the effort. The captains button up their boats for the night, march off to their pickups, and drive home. The fish house owner then plays his role in the charade. He recounts stories about how bad the shrimping has been all summer and curses the agents of the TPWD who, he says, are always snooping around. He closes up the fish house office, turns on the nightlight, and locks the door with a big padlock. He grabs a beer from the cooler and, still cursing, walks over to his mobile trailer standing on concrete blocks behind the fish house and another shed.

For two weeks this play fools me. Each morning I return to the dock between 7:00 and 7:30 to find the shrimpers getting ready to start their work. They return throughout the day to the dock with few shrimp but always in a good humor. They dock their boats after landing only a few pounds of shrimp, drink a few beers, complain about the fishing, and button up their boats again.

Gradually I discover that I am being hoodwinked. A fixture by now on the docks at this small port, I count the number of boxes of shrimp landed by the shrimpers during their day's fishing. Strangely, the number does not match the number of boxes that appear each morning in the tractor-trailer that is periodically hauled off to market. Though some of the extra boxes may come from the handful of shrimpers who fish at night, there are still ten to fifteen or more extra boxes each morning. The fish house operator and the two men who work for him swear that they are closed from 6:00 P.M. to 6:00 A.M. Indeed the operator tells me to come by some night; if I do, he assures me, I will see that the lights are off at the docks, and he will be watching television in his trailer. One morning I count close to thirty boxes, almost 3,000 pounds of shrimp, in the freezer truck. They were not there the previous evening.

The fishermen are successfully evading the law and the anger of other bay shrimpers. Two or three times during the week they return to the dock around 10:00 P.M., and a group of seven to ten captains plans the night's fishing strategy. They fish as a small fleet, dragging their nets in the same part of the bay, pooling information about the location of the shrimp to maximize their time and effort. Sometimes they run without lights to avoid detection. About 4:00 A.M. they stop fishing and motor back to the dock. The fish house owner, in touch with them by CB throughout the night, helps them unload their shrimp without turning on the powerful dockside lights; the unloading, weighing, and loading of the shrimp into the waiting truck are completed in darkness.

It is perfectly legal for the bay shrimpers to fish at night on their bait licenses. But they net more than the 200-pound limit, retain undersized shrimp, and probably trawl in unauthorized waters, strategies that are against the law. The fish house operator and his helpers know what the fishermen are doing and help them any way they can.

Their night fishing done, the bay shrimpers go home for a few hours' rest before returning at daybreak. The schedule is physically demanding, but the men are content because they are making good money from their work; they use this strategy throughout the summer.

I do not believe that this particular kind of illegal shrimp fishing regularly takes place along the Texas coast; rather, it is an exception spawned by a unique set of circumstances. Undoubtedly, a number of other illegal fishing strategies exist that the respondents in the survey chose not to discuss or that I did not observe personally. The diversity of strategies reflects the frustration, anger, and ingenuity of the shrimpers. They believe it is foolish to follow the regulations that govern their fishing. For them, developing illegal fishing strategies is a logical response to irrational policies and laws designed to limit their annual incomes and to put them out of business. The strategies are successful because they are based on the bay shrimpers' expertise and experience and on their commitment to continue as fishermen; they outfox the law enforcers because they have no other choice.

Legal and Illegal Nonfishing Work Strategies

About 25 percent of the shrimpers in the survey work at jobs that have nothing to do with bay shrimping or commercial fishing directly; the men earn an average of $8,000 each year employed in a variety of blue-collar jobs. The fishermen view such employment as a way to continue to shrimp or to bring in money until they can resume fishing. Working night shifts in a convenience store or hiring on at a construction site for a few weeks helps to pay the bills until the bays are once again open to commercial shrimping. In many ways this type of off-boat employment is similar to the practice of small farmers who, along with their wives, may take jobs off the farm in order to keep it going. Often their jobs generate a significant amount of family income.[17] The off-boat workers still identify themselves as bay fishermen. When they work at other jobs, they think of themselves as shrimpers who are temporarily not involved in commercial fishing.

A small number of bay shrimpers are involved in criminal activity that has nothing to do with shrimping. Not considered to be real shrimpers, they are frequently defined by the fishermen as "riffraff," or "no-goods," as "trash, all the family was never anything but trash." The bay shrimpers label them and their families, usually well known in their communities, as deviants. Some are judged as lazy fishermen unwilling to put in the long, hard hours, others as not having the skills or intelligence to make good shrimpers. Regardless of the reasons, they are seen as failures at fishing and consequently at life in general.

Bay shrimpers sometimes joke about how easy it would be for them to resort to smuggling or other illegal activities. They are quite aware of the risk involved, not only to themselves but to their families, and are concerned about their reputations in their communities, knowing they would be tarnished. Even when they are angered by a new law that regulates their fishing more tightly or by a perception of an industrywide trend that lowers their profits, they do not seriously consider involving themselves in the actions of the few professional criminals in the community.

Bay shrimpers are careful to place their own lawbreaking into a category different from the social behavior of community "deviants." Illegal fishing strategies generally are accepted by shrimpers and by the community although some people, especially those elected to public office, may argue otherwise. In contrast, according to bay shrimpers, the individuals who traffick in drugs or who participate in other kinds of illegal activities for a living are nothing more than common criminals who deserve to be caught and punished to the full extent of the law.

4

Why Shrimpers Fail: The Impact of

Imported Shrimp

❖ ❖ ❖ ❖ ❖ ❖ ❖ ❖ ❖ ❖ ❖ ❖ ❖ ❖

Shrimping for many Texas bay fishermen is becoming work they
cannot afford. A minority earn substantial incomes from their efforts, but many more are barely making it. They know they are gradually losing ground, that shrimping provides them a little less each
year than it did the year before.

Some bay shrimpers blame the agents of the Texas Parks and
Wildlife Department, others the fish house operators who buy the
shrimp, their competitors in the Gulf of Mexico, inflation, or the
president. More than a few blame themselves; they reason that if
they could find ways to circumvent the newest fishing laws or get a
higher paying job during the off season, then "things would work
themselves out."

Only a handful of bay shrimpers lay blame on "foreign shrimp."
Fewer than 2 percent of the survey sample ranked shrimp imports to
the United States as the most important problem in the industry
compared to more than 50 percent who rated "existing laws and law
enforcement" first. Fishing laws and dockside price frame the economic and political world of Texas bay shrimpers, determining not
only how and when they fish but also much of what they do after

they dock their boats. Foreign shrimp, imported shrimp from Central and South America and Asia, is rarely designated as a major problem; when it is, the shrimpers label it as "something to watch out for," and "not fair to the American fisherman." Price fixing is another problem that bay shrimpers rarely mention.

Texas bay shrimpers have slipped into the quagmire of a global market economy and do not know it. Increasingly the contemporary macroeconomic trends at the regional, national, and global levels determine the price the shrimpers are paid for their catch at the dock. And there are other economic factors and trends that help or hinder the bay shrimpers and their industry.

The shrimp industry is the most important commercial fishing industry in the United States. In 1991 American shrimpers docked 320.1 million pounds of shrimp, valued at $512.8 million.[1] The Gulf of Mexico regional fishery is the second most important commercial fishery in the United States; shrimp landings are its most significant segment.[2] The five Gulf states landed 228.9 million pounds of shrimp in 1991, 71.5 percent of shrimp landings in the United States, and Texas and Louisiana are the major shrimp producers in the Gulf fishery. Almost 96 million pounds of shrimp were docked in Texas in 1991, 95 million in Louisiana. The other Gulf states landed far less shrimp for this same year: 15 million pounds in Alabama, 12 million in Mississippi, and about 11 million in Florida.[3]

Landings of shrimp netted in Texas bays have increased steadily since 1960.[4] In that year bay shrimpers landed less than 4 million pounds of shrimp, compared to the 1980s when landings ranged from 8 to 15 million pounds. In 1991 13.1 million pounds of shrimp from Texas bays were docked. Annual production from Texas bays averages between 15 and 20 percent of the total shrimp landings for the state; in 1991 bay shrimp accounted for 22 percent of the shrimp landed in Texas.

Bay shrimpers historically have fished for white shrimp, Gulf shrimpers for brown. In the 1960s and 1970s landings of bay white shrimp averaged about 7.0 million pounds per year; during the same period, brown shrimp landings averaged about 3.6 million

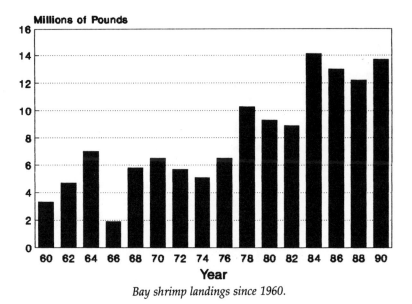

Bay shrimp landings since 1960.

pounds per year.[5] Landings of both species increased, but brown shrimp landings did so at a much faster rate. By 1980 bay brown shrimp landings numbered 8.5 million pounds compared to 6.8 million of white shrimp. Increasingly, bay shrimpers brought in brown shrimp to the docks along with the white shrimp.

Bay fishermen net shrimp that bring a lower value at the dock than the larger ones caught by Gulf shrimpers. In 1991, for example, although bay shrimp constituted 22 percent of the shrimp landings in Texas, they were worth only 13 percent of the value of all shrimp docked.[6] Nevertheless, the annual value of bay shrimp is significant; in 1979, for example, a peak year for bay shrimp landings in the 1970s, the value of Texas bay shrimp was $15.4 million. The Texas bay shrimp industry, when compared to the bay and the Gulf industries in other Gulf states, is among the top two in importance. In 1991 Texas bay shrimp landings (at 22 percent) surpassed the total landings of both Gulf and bay shrimp in Alabama, Mississippi, and Florida.

From 1960 to 1990 Texas Gulf shrimpers consistently landed from 30 to 60 million pounds of shrimp; their harvests have neither decreased nor increased over the last thirty-two years.[7] As in the Texas bay shrimp fishery, short-term trends in the Gulf fishery reflected peaks and valleys that were often extreme from one, two, or three years to the next, but viewed in decades rather than by individual years, shrimp landings in the Texas Gulf fishery have remained remarkably consistent.

Landings of brown shrimp in the Gulf of Mexico have roller-coastered from one year to the next since 1960. From about 60 million pounds in 1960 the figure fell to 30 million in 1961; in 1968 almost 100 million pounds were netted only to drop by 30 million in 1969 and then to 50 million in 1975. Landings were consistently at or above 80 million pounds Gulf-wide from 1975 to 1989, however. Similarly, landings of white shrimp in the Gulf were at the same levels or higher than in previous years from 1975 to 1986; the numbers did fall off somewhat from 1987 to 1989 but were still well within the normal range of fluctuations.[8]

Prices for Texas bay shrimp consistently increased through the 1960s and 1970s. In 1964 Texas bay shrimp sold for $.31 a pound. Prices since that time, given some fluctuation, rose to $1.13 by 1979; from 1964 to 1981 prices increased 177 percent. In the 1980s prices leveled off and then began to fall by 10 to 20 percent or more.[9] Prices for the smaller Texas bay shrimp were consistently lower than for the larger shrimp netted in the Gulf. For example, brown shrimp netted in the bay in 1981 were worth about one-third of the amount shrimp netted in the Gulf brought. Bay white shrimp, though also smaller than the Gulf variety, brought a much higher price but still only about two-thirds of that of the Gulf white shrimp.[10]

A significant gap between the volume of shrimp landings and the price that Texas bay shrimpers received first appeared in 1977 and widened throughout the 1980s and into the 1990s. Texas bay shrimpers were catching more shrimp but were being paid less and less for it. The landmark year of 1986 illustrated the general trend when Texas shrimp landings were the highest in more than fifteen

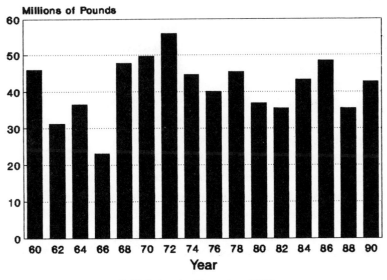

Millions of Pounds

Gulf shrimp landings since 1960.

years, approximately 230 million pounds, an increase over 1985 of 60 million pounds. But the total value of the landings rose only incrementally over the value of the much smaller preceding harvest. Thus, although bay shrimpers netted far greater amounts of shrimp in 1986 than they had for many years, the pay they received was only slightly higher than in the previous year.

While prices in the 1970s and 1980s for bay and Gulf shrimp grew weaker and weaker, demand grew stronger. At both the regional and national levels, sales grew slowly but steadily in the 1970s. Americans consumed 294,000 tons of shrimp in 1970; five years later the figure had increased by about 10,000 tons. From 1980 to 1988 Americans consumed almost double this amount, about 570,000 tons; average annual growth rates in sales for the 1980s were between 7 and 7.5 percent. By the end of the decade, the United States accounted for more than 25 percent of the world's shrimp consumption, followed closely by Japan.[11] Demand in the 1980s was fueled by increased per capita consumption and by population growth. Further, advertising by large retail restaurant chains played an important part in the increased consumption. In the early 1960s per capita consumption was less than a pound of shrimp per person; the rate rose to 2.2 pounds by 1986.[12]

Falling prices and profits did not stop the Gulf shrimp fleet from growing, and the number of shrimp boats and fishermen there increased dramatically in the 1980s. In 1970 there were 8,074 shrimp boats; by 1986 there were 13,041, an increase of 61.5 percent in sixteen years.[13] The reasons for the increase were many, and they varied by region, but the oil boom and bust in Texas and Louisiana coupled with a tightening national economy and lack of alternative employment for those who lived along the Gulf coast were major factors. And there was the promise of nets full of shrimp for the taking. In many parts of the Gulf the myth persisted that shrimping was easy money; the people who tried it soon knew better. In Texas the lure of easy money during the oil boom in the last half of the 1970s, along with the added inducement of Gulf shrimp boats as tax shelters for investors, led to "investor boats," partnerships on paper that brought to the dock new Gulf shrimp boats frequently managed and crewed by inexperienced or marginal fishermen.[14]

The number of Texas bay boats also expanded, mirroring the Gulf shrimp industry until the mid-1980s.[15] Bay shrimp licenses numbered about 1,200 in 1964; during the 1970s the number grew steadily. In that decade bay boat licenses ranged from a low of 3,300 to a high of 4,000, a figure roughly four times that recorded twenty years earlier.

By 1981, when the number of bay licenses reached 5,000, Texas lawmakers decided that the number of bay boats had grown too large, a reaction that stemmed from their concern over the resettlement of Vietnamese along the Texas coast, many of whom had become bay shrimpers.[16] A freeze on new licenses was imposed while they considered various policies; eventually they decided that market forces would thin out the bay boats that were marginal to the fleet. From 1983 through 1989 the number of boat licenses fell sharply; 1987 was an exception, the increase most probably a reflection of the optimism generated by the banner catch in 1986. Only 2,908 bay licenses were sold in 1988 and 1989.[17]

The industrywide trends, when taken together, are puzzling. Landings in the bays increased over time, and the Gulf landings remained stable. The size of the fleets in both fisheries increased; then bay licenses fell off dramatically during the 1980s. Demand for shrimp increased, but the price remained low at the dock. Bay shrimpers were netting more shrimp and receiving less for their catch.

Some industry observers believe that catch per unit of effort (CPUE) explains the economic trends evidenced in the shrimping industry over the last thirty years; CPUE is "the catch taken by a standard unit of gear in a unit of time."[18] Edward F. Klima has reported that the CPUE for both brown and white shrimp from 1960 to 1989 declined, although at different rates. On the average, more shrimp fishermen were each netting fewer brown and white shrimp. At first glance, the CPUE of Gulf shrimpers does seem to clarify certain contradictory data.[19] Landings of shrimp throughout the Gulf of Mexico rose after 1960, then leveled off in the 1970s and 1980s, given the normal fluctuations that were customary. At the same time, the number of boats Gulf-wide substantially increased so that greater numbers of fishermen were netting smaller catches on the average.

There was a decline in brown shrimp landings per boat, most evident in the 1980s if the peak harvest years of 1982 and 1986 are ignored. The average Gulf shrimper netted between 500 and 600 pounds of shrimp per day during this decade; in the 1970s the average shrimper had netted between 600 and 700 pounds of brown shrimp per day. The decline represented a loss of approximately 15 percent of the average catch per shrimp boat from one decade to the next. The data for white shrimp were less clearcut. CPUE did not appear to have declined from 1970 to 1985, but it did fall off substantially for white Gulf shrimp from 1986 to 1989. Given the fluctuations, it is still too soon to know if there is a clear definitive trend in declining CPUE for white shrimp although the evidence to date suggests that possibility.

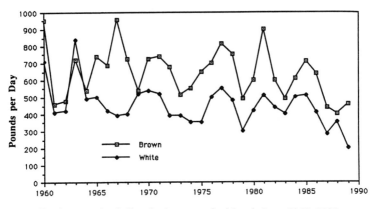

Catch per unit of effort for brown and white shrimp, 1960–1990.

In general, then, Gulf shrimp boats increased in numbers until 1986, but the average Gulf shrimper netted about 15 percent fewer brown shrimp than shrimpers in the 1970s. At the same time, revenue per boat from white shrimp dropped significantly after 1985, but it was not clear if the short-term decline represented a long-term trend. The boat owners whose profit margins were slim took a beating when it came to the size of their catches and the profits they earned at the dock.

No similar CPUE data are available for Texas bay shrimpers. We know, however, that production in the bays rose from the 1960s to the 1980s as the number of boats increased and that it has continued to rise during the 1980s, given expected annual fluctuations. The number of bay boat licenses also rose dramatically during these same years and then fell off by more than 50 percent in 1988 and 1989. Since the number of bay shrimpers declined during these two years, one would logically expect production also to show a decline, given Klima's claims in the Gulf fishery, but this was not the case.

Another factor unexplained by CPUE alone is why so many Gulf and bay boats remain in the fishery. Despite a declining share each year, great numbers of fishermen, in both the bay and Gulf fisheries, still continue to fish, which suggests several possibilities. One explanation might be that some Gulf shrimpers, along with bay shrimpers, are not in the marginal financial position that Klima's data would suggest. Indeed, the data on Texas bay shrimpers support just such a possibility; the burden of declining shares of the shrimp fishery was clearly distributed unequally among bay shrimpers. Based on the annual income data from the survey of bay shrimpers, some earn substantial incomes despite hard times for others.

If accepted at face value, the CPUE argument explains the hard times that bay shrimpers face by emphasizing the ease of entry into the commercial fishery: more and more shrimpers, through aggregate greed, create their own problems as the slice of the pie declines for each. The CPUE analysis ultimately places the blame for the problems that bay shrimpers face squarely on the shoulders of the fishermen. This interpretation not only ignores the long history of laws and policies that directly impact bay shrimpers, but it also overlooks the contradiction of declining prices in the face of rising consumer demand.

❖ ❖ ❖

IMPORTED SHRIMP

The economic impact of imported shrimp provides a much more satisfactory explanation of the financial marginality of many Texas bay shrimpers than does the CPUE analysis. More than thirty years ago Joseph H. Kutkuhn noted that "a 22 percent drop in value despite a moderate increase in yield created economic stress throughout much of the industry. Sharply rising imports were generally credited with having fostered this plight. The situation brightened somewhat in 1960 when the yield rose still higher and its value jumped."[20] Studies in 1969 and again in 1976 also sounded the warning about imported shrimp.[21] Bruce Cox echoed Kutkuhn fourteen years later, observing that "the problem of excessive imports causing high inventories, and in turn low ex-vessel prices, should not be considered an isolated incident but should be recognized as one that could easily occur again."[22]

Kutkuhn and other researchers have been uniformly ignored, however. Demand for American shrimp has steadily increased, but the regional and national shrimp landings, after peaking in the 1980s, have remained constant and prices have substantially declined. Shrimpers blame a variety of factors for hurting their paychecks, including dockside price, but few bay shrimpers or industry experts have taken a serious look at the impact of shrimp imports.

By 1951 the United States was importing large amounts of shrimp, 41 million pounds in that year. Imports continued to grow rapidly, doubling by the end of the decade and increasing throughout the 1960s and 1970s (see Table 4.1). Growth in the 1970s was less spectacular than the previous two decades, but imports doubled again in the 1980s, increasing by 300 million pounds. From 1951 to 1989 the amount of shrimp imported grew to 563.5 million pounds, an increase of 1,354 percent, and it accounted for 70 percent of the shrimp consumed by Americans.

At first glance, the value of shrimp imports also reflects rapid growth, especially evident during the 1980s. For example, in 1965 the value of 170.9 million pounds of imported shrimp was $113.4 million; by 1970 the value had nearly doubled. Prices for imported shrimp more than tripled in the 1970s, from $.81 to $2.65 per pound, even though the poundage of imports was relatively stable. During the 1980s the value increased by almost $1 billion, from $719 million to $1.7 billion. The real price of the imported shrimp, however, increased at a more modest rate from 1965 to 1979, a peak year. If inflation is considered, the price actually declined by about a dollar in the 1980s, showing a steady decrease throughout the decade. In short, it took fifteen years for the price to double from $1.65 to $3.01 (1965–1979). But by the end of the 1980s the price had fallen off to a little more than $2.00, a big decline. Thus from 1965 to 1989 the price for a pound of shrimp increased by forty-two cents.

The astounding increases in shrimp imports in the 1980s were made possible by increased fishing efforts worldwide and by the development of shrimp aquaculture programs in Central America, South America,

TABLE 4.1

Shrimp Imports to the United States, 1951–1989

Year	Amount (million pounds)	Year	Amount (million pounds)
1951	41.6	1971	215.1
1952	38.4	1972	254.5
1953	42.9	1973	230.8
1954	41.4	1974	267.5
1955	53.7	1975	231.5
1956	68.4	1976	271.9
1957	69.5	1977	271.8
1958	84.9	1978	240.4
1959	110.8	1979	269.2
1960	118.3	1980	258.1
1961	128.3	1981	259.1
1962	150.0	1982	319.6
1963	161.6	1983	421.2
1964	162.2	1984	422.3
1965	170.9	1985	452.2
1966	183.6	1986	492.0
1967	202.1	1987	583.0
1968	210.1	1988	598.2
1969	220.1	1989	563.5
1970	247.1		

Sources: William C. Gillespie et al., "An Econometric Analysis of the U.S. Shrimp Industry," *Economics of Marine Resources* 2, Dept. of Agricultural Economics and Rural Sociology, S.C. Agricultural Experiment Station, Clemson University, and Walter R. Keithly et al., "Effects of Shrimp Aquaculture on the U.S. Market," Center for Wetland Resources, LSU, Baton Rouge, 1991.

and Asia. The programs, which varied by country, involved raising higher densities of shrimp than occur in nature by increasing rates of feeding, at the same time carefully monitoring the water quality and guarding against outbreaks of disease. The new shrimp aquaculture was capital intensive; large amounts of human labor were required only in the beginning, when ponds were dug and other infrastructures created. Both private-sector and government funding were used, including the resources of international banks.[23]

Shrimp aquaculture contributed an average of an additional 176.2 million pounds of shrimp to U.S. imports in 1988 and 1989.[24] One consequence of the market glut was much lower prices for American shrimp; without the overseas aquaculture programs, the dockside price for American shrimp in those two years would have risen from $3.27 to $5.44 per pound. Based on these figures, Texas bay shrimpers would have earned as much as 60 percent more.[25]

Latin American and Asian countries radically increased their shrimp exports to the United States in the late 1980s, cornering the lion's share of the world market. China, Ecuador, Thailand, and Indonesia led the other exporters, surpassing countries such as India and Mexico, which had been at the top only a few years earlier (see Table 4.2). For example, the value of Ecuador's shrimp exports to the United States was only $26 million in 1977 but had increased to $362 million by the early 1990s. Thailand increased the value of its shrimp exports by a factor of 7.5 in the same number of years. Traditional shrimp exporting countries found it difficult to compete with the producers who came on line with pond-raised shrimp.

Ecuador and China, among the top new exporters, increased their production through strong cooperative governmental and private-sector capital investment in shrimp mariculture programs, which in most cases were run by scientists schooled in state-of-the-art techniques. Seed stock, water temperature and chemical balance, and food inputs were recorded carefully and when possible controlled and regulated to maximize production. Shrimp were raised in restricted ponds and shallow bays; indigenous fishermen were not allowed to interfere with the growth of the test-tube shrimp. Low-wage laborers were hired seasonally to harvest the shrimp after they reached the desired size.

In the mid-1970s a handful of shrimpers and their professional associations lobbied the U.S. Congress to impose tariffs on imported shrimp but to no avail. Ten years later the shrimpers tried

TABLE 4.2
Countries Exporting Shrimp to the United States, Selected Years

	1977		1986		1991	
	Quantity 1,000 tons	Value $ mil.	Quantity 1,000 tons	Value $ mil.	Quantity 1,000 tons	Value $ mil.
Australia	8	62	13	138	1.3	7.2
Bangladesh	3	17	17	97	10.7	36.5
China	7	52	49	352	77	219
Denmark	12	21	40	150	.1	.4
Ecuador	4	26	31	285	107	362
Greenland	-	-	33	110	.3	1.3
Hong Kong	13	82	31	230	*	*
India	47	178	49	300	38	67
Indonesia	30	138	36	284	25	105
Malaysia	18	53	25	48	7	21
Mexico	30	161	32	361	36	181
Philippines	3	17	11	104	14	57
Taiwan	16	48	51	374	3	13
Thailand	14	57	29	167	100	432

*Figures not listed separately in 1991
Source: "Fisheries of the United States," 1990 and 1992

again; although their efforts were more coordinated and persuasive than the first attempt, the fishermen were unable to convince federal legislators that a tariff would solve the problems of the American shrimping industry. A subsequent government report to Congress pinpointed the American shrimpers' dilemma and the reason that legislators refused to impose a tariff: "There is a significant amount of U.S. investment in foreign shrimp operations, particularly in aquaculture, which export shrimp to the United States."[26] Lawmakers had sided with American interests that had invested in foreign shrimp operations, ignoring the needs of the much less powerful domestic shrimp industry.

Fish house processors, packers, and buyers saw the rise in imports as a boon; they did not join with the fishermen in their attempts to impose a tariff.[27] With a constant flow of imported shrimp, processors could offer large retailers a steady supply. Processors and packers had always suffered from the problem of domestic seasonality; imports meant that they could sell shrimp to retail chain restaurants year-round. Moreover, processors and packers directly benefited from the lower prices imports brought. On the issue of shrimp imports, American fishermen were the losers.

Many Texas Gulf shrimpers made financially marginal by the glut of imported shrimp suffer from other industrywide problems, including overcapitalization. Bay captains who own their own boats, a majority, catch four times less shrimp from their smaller boats than do Gulf shrimpers. But Gulf crews actually earn less than bay boat crews, partly because bay boats require a smaller capital investment, are free of debt sooner, and are much cheaper to maintain. The smaller boats also require less overhead to operate, including fewer crew members. Thus bigger is not necessarily better or more economically efficient. Gulf boat owners in the 1980s required ever-increasing returns on their investments to service the debt, maintain the machinery, and pay the larger number of crew required. Their boats, in comparison to the mosquito fleets of bay shrimpers, are economically less efficient: they run to seventy or eighty feet in length or longer, are air-conditioned, have comfortable kitchen facilities, bunks, showers, and the latest electronic gear, including VCRs for the captain and crew. The owners of such boats must net and dock many more pounds of shrimp to show a profit than bay shrimpers. Thus Gulf shrimpers are much more vulnerable to the vagaries of the shrimp industry; a bad year at the docks can hobble a bay boat owner but finish off a person who has $150,000 or more invested in a Gulf shrimp boat, with its equipment and hardware.

Higher prices would have done much to cover the costs of overcapitalization in the Gulf shrimping industry, but foreign imports stood directly in its path. Texas Gulf shrimpers eventually realized they were outweighed by the combination of powerful American interests and Asian and Latin American shrimp exporters. Gulf shrimpers thus focused on their competitors within the domestic commercial shrimp industry.

Adding to the problems brought on by imported shrimp and overcapitalization is the fact that there are too many shrimpers in the Gulf and in Texas bays. During the 1980s, even as prices fell, more shrimp boats than ever before sailed the Gulf and the bays. A surplus of labor in the Gulf coast states was tied to a regional economy that had gone bust when the price of crude oil fell from more than forty dollars to less than twenty dollars a barrel; the national economy also contributed to higher unemployment rates in the coastal areas from which the labor force of fishermen is traditionally drawn. The number of fishing licenses has never been restricted in the Gulf of Mexico and has been restricted only once among Texas bay shrimpers.[28] For a few hundred dollars any individual can become a shrimp fisherman, regardless of expertise, skills at fishing, or commitment to the work. Though many first-timers fail, some do not, and more fishermen on more boats caught fewer shrimp as the CPUE declined among Gulf of Mexico shrimpers.

Given unrestricted entry into the shrimping industry, stiff tariffs alone on imports will not solve the problem.[29] Restrictions on imports would raise the price of shrimp, but an estimated 1,200 new shrimp boats would soon join the Gulf fleet, according to one researcher.[30] Unlimited entry into the shrimp fishery, both bay and Gulf, remains an important issue that to date has not been addressed.

❖ ❖ ❖

THE SOCIAL IMPACT OF SHRIMP AQUACULTURE AND MARICULTURE

The economic success of foreign shrimp aquaculture and mariculture programs and the potential success of domestic shrimp aquaculture obfuscate a number of crucial questions. If the only measure of the success of shrimp aquaculture and mariculture is productivity, then the figures clearly speak for themselves, demonstrating a thriving industry. Measuring success by productivity alone, however, ignores the fact that the efforts required to generate any new industry take place within the context of local, regional, and national economies and thus have a social impact. Therefore, one must cautiously ask not only how many shrimp are produced but also what social changes occur because of the reallocation of capital and human resources. Ultimately, the crucial questions remain: who directly benefits from the new shrimp industries, and who does not?

Connor Bailey raises several concerns in his work on shrimp aquaculture and mariculture in Asia.[31] For example, he documents the destruction of mangrove forests in Indonesia when shrimp ponds are constructed.[32] Local labor forces help to construct the new ponds and are paid for their work; but once its infrastructure is completed, the new industry requires few men to operate it, only some poorly paid technicians and guards. Loss of the mangrove forests, which serve as a nursery for a variety of local species of fish, including shrimp, translates into the loss of an important and cheap food source for nearby communities. And fishermen who netted and sold the fish to other villages and towns suffer a loss of income.

In Indonesia the use of an important natural resource, the mangrove swamp and the resources connected with it, is reallocated from the poor to local and national elites. Private-sector investors, with help from the Indonesian government, invest heavily in the capital-intensive shrimp farms. Local community elites directly benefit; they serve as go-betweens, pave the way for new ordinances and laws that allow the transfers or use of the land or both to the corporations, and otherwise receive perks and payoffs for their favors.

Bailey also notes that limited financial resources from the government are redirected to building shrimp aquaculture programs, sometimes to the direct detriment of established programs.[33] For instance, he contrasts the benefits of a freshwater fish mariculture program in Indonesia to the shrimp program; direct benefits to the population from the former are immediately lost when governmental funds are redirected. Moreover, there is a loss of local fish as an important source of cheap protein.

The Indonesian government is achieving its major objective, to increase its foreign exchange earnings greatly. Indonesian shrimp are sold for American dollars, which are then used to purchase a variety of foreign goods and services. But Bailey questions the success of the new shrimp industry when it is measured against the real cost in income, food, and disruption of life to the poor.

The Indonesian case is important in and of itself, but it also raises issues relevant to the United States. There is a new and promising shrimp aquaculture industry in Texas. Raising saltwater and freshwater species of shrimp in carefully maintained ponds was heralded in Texas in the early 1980s as the wave of the future. Subsequent research, much of which was undertaken at Texas A & M University, has focused on the reproduction of the seed stock, nursing it through the crucial postlarval stages, and growing it to market size.[34]

Both researchers and investors took as their model the intensive pond programs that flourished in foreign countries; despite their best attempts, the researchers did not succeed. Annual production in Texas has remained under 1 million pounds, less than 1 percent of the total annual production of Texas bay and Gulf shrimpers.

Two major impediments block a successful shrimp aquaculture industry in Texas: shorter growing seasons than in tropical foreign climates and the continuing problem of attaining reliable seed stock at a reasonable price. Species of shrimp have been selected, or developed, that were expected to thrive in the Texas climate. A handful of companies began operations in south Texas, for example, several financed by foreign dollars and guided by technological expertise from abroad.

The efficiency and profitability of the entrepreneurial efforts are problematic; production records are closely guarded secrets, and the financial viability of the start-ups is not yet proven. On the other hand, if shrimp aquaculture programs in foreign countries can skyrocket within a few short years, the same outcome may come about in South Texas. Should shrimp aquaculture prove profitable, investors would quickly jump on the bandwagon, and large amounts of pond-raised shrimp would reach the market in a short time.

The social consequences of Texas shrimp aquaculture have not been considered seriously; the Indonesian case, among others, suggests a cautious approach. If the industry were to come on line with 100 million pounds of shrimp within a two-year period, what would happen to the domestic shrimp industry and its labor force? More shrimp on the market, given the steadily declining prices of the 1980s, would mean lower prices for American shrimpers. Assuming no dramatic upsurge in demand by American or worldwide consumers, American shrimpers would be driven out of business in direct proportion to the number of new aquacultured shrimp that glutted the market.

A large number of coastal communities in Texas rely directly on bay and Gulf shrimpers for their economic livelihood and would find it difficult to replace lost industry dollars, and the bay shrimpers would find a tough job market awaiting them. The most prosperous fishermen would be forced to take big cuts in their annual earnings if they were lucky enough to find a job. In all likelihood, bay shrimpers and their families would be forced to move to a better job market or suffer severe changes in their quality of life.

It is unrealistic to expect that the shrimp mariculture industry, even if it burgeoned, could replace the number of jobs lost to bay and Gulf shrimping. The industry is capital-intensive, not labor-intensive; few new jobs are created after the initial construction of the ponds and other necessary infrastructures; the new jobs are often minimum-wage with few benefits and little chance of promotion.

The impact of a new domestic shrimp aquaculture industry on the Texas environment has received little attention or concern. There has already been a shrimp spill, or series of spills, of undetermined magnitude.[35] In 1990 and 1992 a species of South American shrimp escaped into the Arroyo Colorado River in South Texas; local fishermen reported netting them in the Laguna Madre. Representatives of the bay and Gulf shrimp industries expressed concern over what might happen when the foreign species came into contact with domestic populations of white and brown shrimp; questions on the transfer of disease arose, and many others remained unanswered.

❖ ❖ ❖

PRICE FIXING

Shrimp prices in Texas are suspect.[36] Dockside shrimp prices in the 1980s and 1990s are notable for their uniformity. On a two to five dollar product per pound, depending on the size, the price varied in most Texas ports by less than one or two cents; more rarely the price differential in some Texas ports was as much as five cents. Assuming a three-cents-per-pound difference at four dollars a pound for the larger size shrimp, the price differential was less than 1 percent. As in any other agricultural commodity market, such limited variation strongly suggests that prices are not subject to the forces of a free market.

Texas bay and Gulf shrimpers alike are paid based upon the so-called "Brownsville bid." The price they receive for their shrimp then becomes the basis for the price bid in other Texas ports and in many other ports along the Gulf coast. In turn, smaller processors and dockside fish house operators set their price, squeezed at the top by the large companies and hounded from the bottom by captains and crews who feel they are underpaid or even cheated in their work as fishermen.[37]

From the 1940s to the 1980s ten to twelve fish companies bid for large amounts of shrimp from the six shrimp processors in Brownsville. Each workday, between 1:45 and 2:00 P.M., each company called the processors to bid on their shrimp; bids were made for each size of shrimp available. Each company waited until the last possible minute to make its bid so that at least in theory it could not be underbid. On any given day there might be 2,000 or more boxes of shrimp landed at both Port Isabel and Port Brownsville. The prices bid for the various counts of shrimp became the price for shrimp along the Gulf of Mexico. Whether a shrimper docked his shrimp in Port Lavaca, Texas, or Bayou La Batre, Alabama, the price he was quoted for them was the Brownsville bid or else was based on the Brownsville bid.[38]

The exvessel price for domestic shrimp is still determined by the Brownsville bid, but the bidding process has changed because of the impact of foreign shrimp. There are now only five or six large companies and several smaller local bidders that regularly take part in the Brownsville bid. The amount of shrimp landed at Port Brownsville and Port Isabel has declined substantially; bidders may bid on 300 boxes of shrimp or less, compared with far greater amounts in years past.

Although the shrimp market in Brownsville is much smaller and the number of bidders has declined, the structure of the Brownsville bid has remained constant. Bidders rely to some degree on shrimp prices published biweekly by the trade journal *Seafood Price-Current*, which collects pricing information from a list of its subscribers, including many of the same companies that participate in the Brownsville bid. Prices for different counts of shrimp may vary by a few cents per pound; *Seafood Price-Current* publishes the shrimp prices with a ten-cent-per-pound range.

Bay shrimpers sell their shrimp in a different set of circumstances from that of Gulf shrimpers. The latter either contract their shrimp out annually by volume or sell it on the open market. If a Gulf shrimper sells by volume, he usually contracts with a buyer on a price based on the Brownsville bid with a small additional amount per pound. For example, one Gulf shrimper said that in 1993 he contracted to sell a certain amount of shrimp he had netted for the Brownsville bid, plus ten cents per pound.

Other Gulf shrimpers sell their shrimp on the open market. They dock their catch and ask the fish house operator what his price is for each count of shrimp on that particular day. But the open market is in fact the Brownsville bid. Whether a Gulf shrimper sells his shrimp on consignment or at the dock, the Brownsville bid directly determines the price.

Bay shrimpers have fewer options. Gulf shrimpers often have several different fish house operators to choose from in the larger ports that they frequent; prices may vary by a few pennies per pound. Bay shrimpers, in contrast, usually have only one fish house where they can sell their shrimp. Although they could motor to another port to find a different buyer, in practice they must sell their catch as quickly as possible before it begins to deteriorate in quality. Bay shrimpers are anxious at the dock to unload the shrimp and prepare for another day of fishing; they do not have the time to market the shrimp around, knowing that even if they did find another buyer it would not be worth their while. Prices offered in other ports would be just a few pennies higher, at best, than in their home port.

The particular way in which shrimp are priced, the Brownsville bid, is unique among agricultural commodities. The exvessel price, the result of a few large companies bidding for shrimp docked in Brownsville, is also uncommon in that it determines the market price as the shrimp are processed, brokered, wholesaled, and sent to retail markets.

Increased imports from other countries, often the product of joint ventures by American companies and investors partnered with foreign governments and private capital, also may have changed the ways in which prices of foreign and domestic shrimp are established.[39] Processors have benefited tremendously from a regular supply of foreign shrimp since they are no longer dependent on the seasonality of domestic shrimp. Availability has allowed supermarket chains and food services the opportunity to create new markets, increasing demand. Shrimp is now sold year-round in many chain groceries, along with other fresh fish. Retail restaurant chains, supplied by food service companies, are able to offer shrimp as a permanent item on their menus. Processors have grown in proportion to the availability of foreign shrimp to the American market, becoming the most important link in the organizational structure of the market.

At various levels, kickbacks and price fixing are a part of marketing shrimp in the 1990s.[40] Just as with price fixing, the economic impact of kickbacks between brokers and wholesalers falls on bay and Gulf shrimpers. And kickbacks ultimately may mean that producers receive lower prices for their shrimp and that consumers, at the other end of the market, pay higher prices.

❖ ❖ ❖

A GLOBAL SHRIMP MARKET

Although the traditional practice of price fixing has not disappeared along the Texas coast, the major problem that bay shrimpers face is an oversupply of shrimp available on the American market. Too much shrimp is produced in foreign countries, and it competes successfully with American shrimp. It is a buyer's market. Regardless of how hard he works on the water, a bay shrimper still must confront a low price at the dock for his catch. Many shrimpers blame themselves for their declining incomes and failing industry. Still, their only option on an individual level is to develop new fishing strategies that will increase their catch; that is exactly what most bay shrimpers do.

The fishermen, then, are not becoming economically marginalized because of flaws in how they do their work or because of personal shortcomings. They are failing because their industry as a whole is failing, because Indonesian and Chinese peasants can produce cheaper shrimp, given present domestic policies and regulations governing bay shrimping. And many of the operations abroad are the result of entrepreneurial efforts stimulated by American capital.

The complete story of bay shrimpers, however, must go well beyond pointing the finger at broad economic trends. Macroeconomic forces, such as foreign competition in the controlled production, harvesting, and marketing of shrimp, are played out within international and domestic political arenas. The role of local, state, and federal governments in the bay shrimping industry must be examined. These levels of American government have traditionally supported bay shrimpers and consistently intervened on behalf of the fishermen's best interests; the government is no longer carrying on that tradition.

5

The Role of the State: Government

Regulation, Unionism, and Shrimpers

❖ ❖ ❖ ❖ ❖ ❖ ❖ ❖ ❖ ❖ ❖ ❖ ❖ ❖

The relationship between bay shrimpers and the individuals who regulate fishing in the 1990s has developed over a long period. Although the bay industry boomed in the 1920s, its history is rooted in the past; government has played at least some role in the fishermen's lives from the beginning of Texas statehood. From the 1920s to the 1950s regulations began to define the work of commercial shrimpers on the water and at the same time to protect them from competitors. State government in the 1950s, however, shifted its position so that the bay shrimpers perceived it as the enemy; in the 1970s the federal government joined forces with the state. Many bay shrimpers still find it difficult to accept that their government, including local officials, is not on their side.

❖ ❖ ❖

EUROPEAN SETTLEMENT

Bay shrimping as first practiced by European colonists along the Texas coast was a means of subsistence, an activity that sustained the household when other foods were unavailable or scarce. Early settlers netted in season a rich variety of fish from the nearby bays and rivers, which, along with local game, fed them in good and bad crop years.

In fits and starts settlers colonized the upper third of what is now the Texas coast. Stephen F. Austin's original colony, begun in 1822, was a part of Mexican Territory and lay between the Brazos and the Colorado rivers, at some distance from the coast.[1] He planned the colony carefully in order to avoid unnecessary battles with the indigenous Indians, especially the Karankawa, hunters and gatherers who for hundreds of years had ranged along the coast of the Gulf from what is now Texas to deep into Mexico.[2]

Small ports along the coast sprang up to import goods to supply the new colonists and eventually to export their products to the outside world. In these small communities residents often depended on the fish they caught from the rivers, the bays, the extensive estuaries, and the ocean surf as an important part of their diet. As the Karankawa had done before them, the new immigrants discovered many species of fish that could be caught by line or by casting simple handheld nets: trout, redfish, flounder, sheepshead, saltwater catfish, and many others no longer abundant today. The colonists also found shrimp everywhere in the bays and rivers and surf. They were easy to net by hand; children could catch them.

Matagorda was developed as a part of Austin's settlement of Mexican Territory in 1829. The first European colonists, arriving there aboard the ship *Little Zoe*, had difficulty making land and for several days were becalmed in Matagorda Bay. After landing, they soon met with a small band of Karankawas, who, according to the colonists, "manifested no hostility. Their canoes were stored with fish, all neatly dressed, which they bartered to us."[3] But later the Karankawa murdered some colonists, the settlers retaliated, and before long killings between both groups were common news. A log fort was constructed in Matagorda to protect the colonists, and by 1831 town lots were sold for thirty dollars apiece.[4] Settlers began streaming in, some choosing to live in Matagorda, most bound for Austin's colonies in the interior.

Port Lavaca was established by the Mexican government around 1815 to supply their forts at Bexar, La Bahia, and Nacogdoches and their forts-to-be on the Brazos, Galveston Bay, and the Neches and Lavaca rivers.[5] Modest port facilities were first built at Port Lavaca in the late 1830s. Corpus Christi began as a trading post, established by Henry L. Kinney in 1839, on territory claimed by both Texas and Mexico; Kinney fought off Mexican bandits, outlaw Anglos, and Indians.[6]

The first Texas census of 1850, initiated after its independence was declared, counted 31,850 citizens in the fifteen existing coastal counties.[7] Cameron County, on the extreme southern tip of the Texas coast, accounted for almost one-third of the number; the town of Brownsville on the Rio Grande having the majority of the county's population. The tiny Cameron County ranching community of Point Isabel (eventually renamed Port Isabel), on the Laguna Madre, had only a few hundred residents.[8] Harris County, which fronts part of Galveston Bay, was already home to the town of Houston, but in 1850 the county boasted less than 5,000 people; Galveston County had slightly fewer residents. About 4,800 people were living in Brazoria; the other coastal counties had around 2,000 or less. Refugio and San Patricio counties, for example, listed 288 and 200 people in the 1850 census.

Federal policy regarding American fisheries was minimal until the creation of the office of Commissioner of Fish and Fisheries in 1871. A century before, after winning independence from the British, the new American government initially had taken a strong protectionist view of its fisheries. American fish exports were subsidized, as was the building of larger American fishing vessels, and imports were taxed. The specific right to regulate American fisheries and to enforce the regulations was not delineated in the Constitution but came under the broad powers of the new nation to conduct free trade with other countries without interference from the British.[9]

It was in this political context that individual states claimed control of local fisheries through judicial rulings. For example, the oyster industry in the newly independent United States was one of its most important fisheries. A series of legal cases formed the basis of an individual state's claim both to the access of oysters by fishermen and to the total regulation of that fishery. By legal reasoning, the precedent established for the oyster industry was applicable to other fisheries, including shrimp.

Beginning with *Arnold v. Mundy* in 1821, the supreme court of New Jersey successfully defended the state's right to the oyster grounds off its immediate shores and the authority to regulate the oyster fishery. The state argued that the fishery resource was a public trust and that the state should hold ownership of that trust in the best interests of the citizens of New Jersey.[10] In 1842 the U.S. Supreme Court agreed, further arguing that states had the right to keep out-of-state residents from freely using the oyster fishery.[11]

Federal interest in commercial fisheries was renewed soon after by the first meeting of the American Fish Culturists Association in 1870; only a year later Congress was persuaded by the organization to establish the office of the U.S. Commissioner of Fish and Fisheries. Its immediate tasks were to determine which fishery resources had been depleted by the American fleets and to develop practical remedies to the problems it discovered. The commissioner was given no regulatory powers, but Congress did give him a $15,000 budget in 1872.

In concert with a majority of the large fish corporations that dominated the commercial fishing industry, the commissioner strongly believed that artificial propagation of commercial fish was the only solution to the problem of depleted fish stocks.[12] The approach meant that the federal government would not protect certain species of fish by limiting access to them; it would simply find ways to grow more fish. Such views found unanimous support from the American Fish Culturists Association (later the American Fisheries Society), a group of natural scientists who quickly developed the philosophy, logistics, and rhetoric of fishery management.

Texas, gaining statehood in 1845, at first displayed little interest in fish cultivation, but its legislature did show a concern, if temporary, for depleted fishery stocks and passed a bill in 1879 that encouraged new ways to preserve indigenous species by erecting fish ladders to help them reach their spawning grounds.[13] The office of the Texas State Fish Commission was soon created; the commissioner had as his major duty to oversee the construction and maintenance of fish ladders on the state's primary river systems. But no budget for the commissioner's salary or for operation of the office was forthcoming until the legislature met in 1881. It allotted $5,000 per year for two years, and in 1883 the legislature met again and provided $10,400 per annum.

By the time the office was finally funded, the state legislature had fallen into line with the federal government's decision to focus on the replenishment of fishery stocks, to grow more fish rather than to protect them. The Texas commissioner was mandated to cultivate and distribute fish throughout the state's rivers and freshwater ponds. By 1885, however, the legislature lost interest in funding the state office; cultivation and stocking ceased and the office was abolished.[14]

Even though Texas state government lost heart in the task of replenishing fishery stocks, the federal government continued the work, and the U.S. Fish Commission stocked "young scad, salmon, rainbow trout, carp" from 1880 to 1884. The carp and trout were a success, the other stock a failure.[15]

Again in 1890 the U.S. Fish Commission showed its interest in Texas when it sent 745 small lobsters to Galveston to be released into the Gulf of Mexico. The experiment was judged a failure when only a few of the hundreds of lobsters were netted by local fishermen. In the early 1890s the Fish Commission also established a federal hatchery in Texas for the purpose of replenishing depleted fishery stocks.[16]

❖ ❖ ❖

THE SHRIMP FISHERY: 1870 TO 1920

There was little concern either at the state over the national level for Texas shrimp; the state's commercial fishery was simply too insignificant to regulate during this period.[17] The powers of the free market were assumed to be in control, possible state or federal policies and regulations, or both, an infringement on the natural economic forces of supply and demand. The one exception was the Galveston Bay fishery, where fears grew after a decline in production.

A comprehensive overview of the nascent Texas coastal fishing industries, including the commercial shrimp fishery, was undertaken in 1891, the first of its kind. Charles H. Stevenson, a fishery agent for the U.S. Fish Commission, canvassed the Texas coast, documenting the status of the fisheries in every coastal community.[18] Stevenson's first-person account described a tiny Texas fishing industry driven by local markets and worked by men who fished a variety of different species in season. Although they worked for a meager wage, bay shrimping was a job for a small group of fishermen; for most other people who fished along the coast, shrimping remained a subsistence activity, a way to put food on the table at no cost except for their time.

Markets for the fish harvested by the small labor force were severely limited because there was no way to transport the product to major population centers. The one exception was Galveston, a city of more than 30,000 and home to the most important but still modest shrimp industry and to a thriving oyster industry. Shrimp, along with other fresh fish, were sold at open-air markets and door-to-door in Galveston by enterprising fishermen and their wives to earn extra household income. The wealthy of the city ate shrimp as a side dish to their main meat courses, but the poor consumed it because they could buy it inexpensively or fish for it themselves from the bay and the Gulf surf. They sometimes called it "cheap fish."[19]

The other fishing communities and ports spanning the Texas coastline were at a great distance from potential markets, and the warm and humid climate made it extremely difficult to preserve the catch. Lacking rail lines or other reliable methods of transportation from the coast to more densely populated Texas urban areas in the interior of the state, the abundant fisheries were of little economic

Bay boats docked at Galveston.

value. For lack of viable markets, large catches had to be given away, or they simply rotted at the dock.[20]

The work of the Texas fishermen was "very rough."[21] Bay seining, the dominant fishery besides oystering, was accomplished from wooden, shallow draft vessels powered by sail; the boats were rarely larger than five tons. There were no major boat builders in Texas, and the majority of the boats were constructed by the fishermen themselves.

Fish, including shrimp, crab, and sea turtles, were netted in the bays or close to the shores of the barrier islands by crews of two to four men. Most commonly the seine was spread from the aft of the boat into the bay until it reached its full length; then several men laboriously hauled in the large nets. If the sailboat was fishing the bay, it was gradually maneuvered into shallower waters. A fisherman with leather boots then jumped into the bay and carefully removed all the stingrays before the other crew joined him.[22] The fishermen continued their backbreaking work, collecting the net and culling through the catch. The most marketable fish—redfish, trout, and sheepshead—were immediately thrown onto "live-fish cars," shallow skiffs constructed of rough planks. In descending order, reflecting the price received at the dock, were "croakers, jackfish, hogfish, drum, mullet, bluefish, Spanish mackerel, pompano, rockfish, jewfish, pigfish, and whiting."[23] The live-fish cars held from 400 to 2,000 pounds of fish and were towed behind the sailboat directly to market. Texas commercial fishermen sometimes fished close to the shores of the barrier islands and on occasion ventured into deeper Gulf waters for red snapper. Some Texas vessels were caught in storms while fishing offshore and never returned to the dock.[24]

Bay seine fishermen were paid using the share system.[25] The captain and crew received shares, the captain's share no greater than the rest; the boat owner also received one share. The price for fish by the 1890s was set for the season and did not change. Earlier on the fish house owners had sometimes owned boats and lent them to the fishermen in return for a share. But Stevenson reported that "the practice has been discontinued to a considerable extent, as the fisherman fail to take the best care of the boats and seines when they have no property interest in them."[26]

Texas fishermen averaged wages of about $325 per year by 1890, a slight income by standards of the day. To supplement their annual earnings, commercial fishermen engaged in a wide variety of other activities including, "hunting and marketing ducks, geese, and other food or plumage-bearing birds with which the bays along the Texas coast abound during certain seasons of the year."[27]

Commercial fishing in Texas was notoriously unreliable from its earliest beginnings. The different bay systems provided fish for the seiner's nets only during certain times of the year. One bay might be productive several years running, only to be much less so the following year. Moreover, the annual seasonality of fish in Texas bays determined their availability from one year to the next. As a result, many Texas commercial fishermen, even though full-timers, were forced to work in other fisheries or at nonfishing jobs.

Bay boats tied up at the docks at Matagorda, Texas.

By 1890 some bay seine fishermen regularly migrated from Galveston Bay to Matagorda Bay during slack times of the year and thus were able to maintain their work as fishermen in spite of seasonality or other factors constraining the fishery. The fishermen sent their catch to market by way of steamer, which sailed from Port Lavaca to Galveston.[28]

Fish that had little or no market value in Texas, cheap fish, were saved by commercial fishermen for their own use and were an important source of food for them and their families. There was a large local demand among the poor of coastal Texas for cheap fish although the price they could pay was only a few pennies.[29] White shrimp were caught in the bays year-round and were preferred by far over brown shrimp, which were seasonal.

Prior to 1890 the most important commercial fisheries in Texas were Galveston and Matagorda bays; the other large bay systems, including San Antonio, Aransas, Corpus Christi, Espiritu Santu, and the Laguna Madre, were of lesser significance. Along these extensive bays and their secondary bay and river systems—Copano Bay and Mesquite Bay were notable examples—lay a small number of tiny fishing villages and small communities, some the original ports founded by Austin's colonists or by Mexicans. By 1890 Galveston County had a population of 31,476, which accounted for about one-fifth of the total coastal-county population in Texas,. The population of other coastal counties was much smaller: Brazoria County had 11,506 citizens; Jefferson, 5,857; Matagorda, 3,985; Nueces, 8,093; and Orange, 4,770. Calhoun County, home of Port Lavaca, the once-thriving town established by Austin's settlers, had suffered badly after the Civil War and had only 815 residents.[30]

Galveston Bay was the only significant commercial shrimp fishery in Texas in 1890. Ten small boats, each with crews of two fishermen, regularly fished its waters.[31] A small shrimp cannery had been built there in 1879 and in 1880 produced 75,000 pounds of shrimp in one-pound cans.[32] It operated for only a few years before closing down.[33] Shrimpers fished large seines similar to those used by other Texas commercial fishermen but of a much smaller mesh size than for trout or redfish.

By the time of Stevenson's first report, the fishery experts believed that Galveston Bay was overfished by local shrimpers. Production of shrimp had declined and by 1890 was at 9,200 buckets, fifteen pounds of shrimp to each bucket. It is highly unlikely, however, that a small fleet of ten boats caused a major decline of shrimp in Galveston Bay, given the crude means by which shrimp were harvested, the number of boats involved, and the ability of shrimp to reproduce from one year to another in large numbers. Other causal factors almost certainly led to the decline in production, including pollution of the bay waters. Production rebounded a decade later, however. Galveston Bay shrimp were judged inferior in size and compared unfavorably with shrimp netted in Louisiana or other Gulf states. Although the future looked bright for the Texas oyster industry, the outlook for the Texas commercial shrimp fishery was not promising.[34]

One particularly intriguing type of commercial fishing was the "wagon fishery" of Galveston Island. Most full-time Texas fishermen sailed back to shore with their catch and sold it as soon as it was offloaded at the dock, but one group hauled their fish to market in their farm wagons. In contrast to most commercial fishermen of the day, this small group, who fished with surf seines during certain seasons of the year, were primarily truck farmers, not full-time fishermen. Most Texas fishermen by 1890 fished full-time, but a minority fished as a part-time activity to earn extra income whenever possible. The wagon fishermen of Galveston Island spent most of the year farming on Bolivar Island and on portions of Galveston Island as their major source of livelihood.

Women were a significant part of the wagon fishery and the other commercial fisheries in Texas. The wives of the wagon fishermen often sold fish door-to-door in Galveston, gaining a better price for their product than they could at market.[35] The practice spread, eventually, along the Texas coast.[36] Although commercial fishing in Texas was exclusively a male activity, women played a vital role in the distribution and sale of the catch. Both male and female fish peddlers were a common sight not only in Galveston but also in Corpus Christi and Port Lavaca.

Shrimp were netted commercially during this period in Matagorda and Corpus Christi bays and the Laguna Madre, but annual production figures were quite small, rarely exceeding a few thousand pounds a year. Small commercial fisheries for other species, including redfish, trout, and sheepshead, were operating along the coast south of Galveston, and in Matagorda and Corpus Christi bays a viable oyster industry had also been established.[37]

From 1890 to the early 1920s commercial shrimping in Texas stalled. For example, in 1918, a typical year, a total of only 164,067 pounds of shrimp was docked, a figure insignificantly higher than the one recorded in 1890.[38] Shrimp production shifted from Galveston Bay, where about 32,000 pounds of shrimp were netted in 1918, to Corpus Christi Bay, where 145,000 pounds were caught the same year. Shrimp production was negligible throughout the rest of the

Texas bay and estuarine systems with the exception of the Sabine Lake area, where almost 41,000 pounds of shrimp were landed; in Matagorda Bay, for instance, less than 600 pounds were netted.[39]

Palacios typified the tiny fishing villages dotting the Texas coast; it had only three fish and oyster houses in the early 1900s. One was the Ruthven Packing Company, founded by A. R. Hillyer and Duncan Ruthven; the owners hauled their catch by wagon to a nearby warehouse where it was iced and packed for shipment.[40] The other two fish houses were Deutch's Liberty Fish and Oyster Company and J. J. Burke's Fish and Oyster Company. The fishing boats, powered by sail, included the *White Wing*, the *J. T. Hicks*, the *Falcon*, the *Sterling*, and the *Red Wing*. They raced each other every Fourth of July in front of the town pavilion, which faced the small bay.[41]

A representative of the Palacios Development Corporation convinced an Alabama shrimper, Fred Bates, Sr., to visit the town in 1918. The next year Bates returned with two boats, the *Helen B.* and the *Emma*, and the Alabamians netted all the shrimp they wanted but could find no market for their product. Several years later they returned: "They caught 1,200 pounds of shrimp in one short drag with a 35 foot net. 600 pounds of the catch were shipped to Beaumont, but the rest could not be sold. Word was passed around town that anyone wanting shrimp should bring their containers and fill them. The remaining shrimp were shoveled overboard. The Bates family did not try shrimping for a few more years."[42]

Although the Texas shrimp industry was languishing, the commercial fish industry evidenced some growth, particularly in the harvesting, canning, and distribution of oysters, and the work force of commercial fishermen grew. In 1880 the number of full-time fishermen in Texas was 291, with an additional 200 employed part-time throughout the year; a decade later the numbers were 819 and 286.[43] By 1918 the number of commercial fishermen, including oystermen, was 1,427.[44] Trout and redfish were still in modest demand, and a small red snapper industry also developed. Oysters, however, continued to dominate Texas fishing; almost 3.5 million pounds were harvested in 1918, valued at $143,610.

Several important changes occurred in the infrastructure of the industry and in fishing technologies to create a boom in Texas bay shrimping during the 1920s. Among the advances were the linking of Texas ports by rail to the state's interior, improvements in ice and freezer technologies, the development and promotion of the otter trawl by the federal government, and increased knowledge of more efficient cannery techniques.[45]

The town fathers of several small Texas ports struggled for many years to bring the railroad lines to their docks; some community leaders finally achieved their goal but only after extensive efforts. Typical among these individuals were entrepreneurs in the town of Aransas Pass.[46] The dream of these businessmen was to deepen and widen the pass that flowed through the barrier island to the Gulf of Mexico; later they planned to develop their harbor facilities and turn Aransas Pass into another New Orleans. The key to their ambitious dreams was a railroad line. Like most small Texas ports on the Gulf of Mexico, Aransas Pass was served several times weekly by the Morgan line of steamers. Attempts to build various tracks that would tie the small port to San Antonio, Corpus Christi, and other larger population centers to the north met with repeated failures.

A major part of the problem for Aransas Pass was competing rail interests. The Southern Pacific Railroad promoted Galveston as its port of choice for the Texas coast, routing all its trade through that city. The company actively discouraged any independent railroads from building lines to other Texas coastal ports.

Colonel Uriah Lott financed and completed a line linking Aransas Pass to San Antonio and points south in 1886. The community was justifiably proud of the considerable accomplishment; rail lines had to be built through miles of low-lying wetlands and bays. Lott supervised the difficult work, refurbished old railroad cars and engines, and formed the San Antonio–Aransas Pass Railroad (SAAP).

Fishermen from Matagorda Bay moved to Aransas Pass, and two fish houses were built as soon as SAAP's last rails were in place. By 1890, 210 men were engaged in the fishery, which included oysters and green turtles.[47] The celebration of the new transportation link was short-lived, however, as SAAP soon went into receivership, eventually landing in the hands of Capt. Mifflin Kennedy, one of the founders of the King Ranch. The Southern Pacific Railroad acquired the line and for many years ran it in a halfhearted manner, preferring to route the majority of its rolling stock through the port of Galveston as was its policy. Aransas Pass remained a small Texas coastal community, its potential untapped.

Port Lavaca and Freeport, two of the largest fishing communities, acquired rail lines toward the turn of the century, but there was only a limited market for their shrimp, fish, and oysters; fish had to be delivered within the day to consumers or risk spoilage. Canneries were one obvious solution to the problem, but entrepreneurial capital was required for start-up costs, along with a guaranteed supply of fish to keep the labor force busy at the machines. Canneries along the coast of Texas were problematic, as evidenced in Galveston. As the Texas oyster industry developed, however, canneries were constructed once again on Galveston Island and in a few other coastal communities.

Gradually, the technology to produce large quantities of ice at relatively low costs was developed and refined. The knowledge of how to freeze, handle, package, and distribute frozen shrimp, however, was much slower in coming, and the process was much more expensive than running oyster or shrimp canneries. As late as the 1930s, 96 percent of the shrimp harvested in the Gulf and in the South Atlantic shrimp fishery was either delivered fresh to market or canned.[48] Problems in packing houses and canneries stubbornly persisted: sometimes the finished product, a tin of shrimp, was not processed according to uniform standards, an error that had "a great adverse effect on the consumer popularity of canned shrimp."[49]

The relatively crude method of harvesting shrimp was another major impediment to the development of a viable industry. Texas shrimpers, along with their fishing neighbors in Louisiana, Mississippi, Alabama, and Florida, continued to use large seines to capture the crustaceans, backbreaking work that required endless hours of laying out and hauling in nets by hand.

In 1912 and again in 1914 the Bureau of Fisheries used small otter trawls to collect specimens at Beaufort, North Carolina.[50] It is unclear whether the bureau was intent on demonstrating the new net to local commercial shrimpers or if it had some other purpose in mind. At any rate, local shrimpers observed the use of the net by the federal regulatory agency, a net similar to the configuration still in use in the 1990s. The new otter trawl was easier to use and more efficient; shrimpers who used it caught more shrimp. Adoption of the new net was remarkably quick, and by 1917 most shrimpers in Texas and in the rest of the Gulf states were fishing with it.

The otter trawl had other consequences as well. The net allowed for better fishing in deeper waters, both inside the bays that were sheltered by the barrier islands and along the rivers and their mouths that fed into the Gulf of Mexico. The new trawl also opened up fishing grounds farther off the barrier islands, in the deeper waters of the Gulf that largely had been untapped by shrimpers.

Demand for Texas shrimp still remained problematic. Established markets in Mississippi, Louisiana, and Florida were too distant for most fisheries south of Galveston Bay. The local and regional markets remained stagnant; the population of coastal Texas showed only modest growth as a percentage of the total state population.[51] By 1920 the coastal population had increased from 7 percent to 10.5 percent of the total state population, but it was not equally dispersed. Almost half of the 489,735 residents along the coast lived either in Harris County, synonymous with Houston, or in Galveston County. Nueces County, for example, home of Corpus Christi, numbered only 22,807 people in the 1920 census.

Another problem tied to demand was that shrimp had a bad image; it was cheap fish, readily consumed by lower income groups. There was only a small market for white shrimp; no real market existed for brown shrimp. Even in the 1930s promoters complained about image problems. Affluent Americans who ate white shrimp liked it primarily as an extra ingredient in salads; it was not considered a meal in itself, a legitimate main course. The solution was "national advertising to promote the wider use of shrimp. This should be especially concentrated in northern areas of this country and should forcefully point out the many preparations possible from shrimp other than the salad so frequently associated with this crustacean."[52]

With most of the problems at least partially if not satisfactorily addressed, the Texas bay shrimp industry boomed in 1923. In 1918, 164,067 pounds of white shrimp were harvested from Texas bays; five years later the catch was 3.5 million pounds.[53] Shrimp production in Texas continued to expand, tripling by 1927 when it reached a peak for the decade of almost 12 million pounds. Through the next two years production declined somewhat, presumably as a result of normal variations in the shrimp population. Then in 1930 the number jumped to 10 million pounds; in 1931, 14 million pounds were harvested by Texas bay shrimpers.

By the end of the 1920s bay shrimping in Texas was a legitimate if still modest industry, with the value of the shrimp docked estimated at $400,000. Growth in the industry, when it finally came, was rapid and strong. The number of men who netted shrimp from boats and with seine nets from shore numbered 488 in 1927 but more than 600 by 1930.[54] The number of boats weighing less than five tons, the favored size for bay shrimpers, is estimated at 192, a significant increase in a brief span of years. There were twenty-three shrimp boats in 1927 greater than five but less than ten tons, and five boats greater than eleven but less than twenty tons.[55] There were an estimated 237 Texas bay boats by 1931.[56] The larger vessels were powered by gasoline motors, but only rarely were diesels used. The majority of the smaller sailing vessels were gradually converted to gasoline-powered boats after 1926, and by the mid-1930s sailboats were no longer used in the Texas shrimp fishery although they remained common south of the Rio Grande.[57] In that decade Texas shrimpers also adapted small gasoline engines to their work decks, using them to drive winches, a rather simple technological transfer that greatly facilitated the retrieval of the otter trawls.

In tiny Palacios, Carlton and John W. Crawford opened a plant to can figs, chili, and tamales, and in 1930 they traveled to Alabama where they persuaded several shrimpers to move to their community to net shrimp for the cannery. The Alabamians brought their shrimp boats with them and built new ones in Palacios. H. L. Lantron built another fish and oyster cannery there so that during the 1920s the town had four fish houses to serve the oystermen and the growing number of local commercial fishermen who turned to shrimping.[58]

There were over forty wholesale fish houses in Texas in 1931, a sizable increase from the twelve that existed in 1890. Galveston and Corpus Christi became the centers of the commercial activity, but the small communities of Port O'Connor, Port Isabel, Port Lavaca, Palacios, Rockport, Ingleside, and Port Aransas were also important.[59]

The fish houses employed laborers who worked for a piece-rate wage. The majority were women, including a significant number of Mexican Americans, blacks, and undocumented Mexican workers; they became a crucial element in the bay shrimp industry, especially the minority women. Children also were commonly employed during peak seasons in the fish houses.[60]

The growth of the bay shrimp industry in the 1920s and 1930s was not reflected by the number of canneries in Texas. There were two in 1923 and three more the following year, but by 1932 the number had not increased. Shrimp canneries remained centered in Mississippi and Louisiana; at one time during the 1920s Mississippi was home to twenty-eight shrimp canneries, Louisiana to thirty, although by 1931 the number had leveled off to fifteen in Mississippi and seventeen in Louisiana.[61] Shrimp canneries were capital-intensive, relatively risky ventures, given the failure of canneries before 1900. Their small number in Texas during these years places the state's shrimp boom in perspective; bay shrimp netted in 1923 accounted for only 4.8 percent of the total shrimp production in the Gulf states.

Alabama's shrimp production closely resembled that of Texas, but elsewhere along the Gulf of Mexico substantial increases in annual production after the early 1900s dwarfed Texas' production.[62] Florida fishermen netted more than 3 million pounds of shrimp as early as 1902 and in 1923 docked almost 14 million pounds. In Mississippi production was substantial; for example, by 1890 it exceeded 1 million pounds and was at 10 million when Texas shrimping finally boomed in 1923.

Louisiana, however, was at the forefront of the industry, the major producer of shrimp from the Gulf of Mexico. In 1887 about 7 million pounds were docked in that state, and by 1923 production was a 28 million pounds.[63] Although significant, the boom in Texas shrimping was modest by the standards of the day.

Both the timing of the boom in Texas (similar industries in other states, with the exception of Alabama, were already established) and the absence of canneries helped determine the fate of bay shrimpers there.[64] Local unions of bay shrimpers did not form in the 1920s and 1930s as they did in other Gulf states because there was little reason for them to form in Texas. When bay shrimpers finally struggled to organize, the courts outlawed the unions.

Strong and viable shrimp industries in Louisiana, Mississippi, and Florida kept their fishermen at home. The Texas fishery was particularly unattractive to outsiders because it was relatively underdeveloped, strung out along a vast and unfamiliar coastline stretching to Mexico. Moreover, the shortage of shrimp canneries made transporting Texas shrimp to factories hundreds of miles away an extra cost that non-Texas shrimpers did not want to pay. Thus states bordering the Gulf of Mexico left Texas shrimp, for the time being, to Texas fishermen.

Meanwhile, shrimpers in Louisiana and Mississippi had their own problems, which soon had a major impact on Texas bay shrimpers. The canneries and the processors in the two states initially controlled the fishermen and the fishery. But shrimpers struggled to receive higher prices from the canneries, to raise and stabilize their wages in order to provide a better life for themselves and their families. Further, they were facing increasing competition from out-of-state shrimpers.

Before the 1930s the majority of canneries and processors in Louisiana and Mississippi owned their own shrimping vessels.[65] Unlike the boats in the Texas fishery, the sailing schooners were large vessels that required crews of ten or more to operate and maintain them. As late as 1927, for example, there were still twenty-eight sailing schooners in Mississippi, nineteen of which were larger than ten tons. Cannery owners with their own boats were thus able to greatly reduce the financial risk of operating their factories. Independent shrimpers, such as those in Texas who owned their own boats, could always refuse to sell shrimp to the canneries; their political clout lay in the potential to withhold shrimp, to idle the expensive machinery. But cannery owners who owned their own fleets stabilized the availability of shrimp because they could dictate to the shrimpers who worked for them when to fish, how to fish, and the wage they would be paid. The shrimp fishermen were powerless, given the economic arrangements.

With the advent of the gasoline motor, the conditions gradually changed. Large sailing schooners were replaced by much smaller boats that were built, owned, and operated by commercial fishermen not employed by the canneries. In Mississippi, for example, by 1927 there were still twenty-eight sailing schooners crewed by about 108 shrimpers.[66] Yet there were also 265 motorized boats of less than five tons crewed by 530 men.[67] The Mississippi and Louisiana canneries lost their control of the shrimp industry when sailing schooners became outmoded, and shrimpers, in turn, gained power. By 1928 only one sailing schooner remained in Louisiana.[68] Shrimpers demanded a higher price for their catch, converting the increase into a higher annual wage and an improved quality of life. They also sought to protect their own state fisheries from outsiders, in effect to limit entry through the power of the union.

Fishermen's unions formed in the 1930s in most of the port cities of Mississippi, Louisiana, Alabama, and Florida.[69] The unions sought to negotiate higher prices for shrimp for their members, working with the trade associations to which the canners and packers belonged. They also developed a number of other strategies to gain control over the fishery resource, thereby minimizing the power of the factory owners. The strategies included price fixing, a rudimentary attempt to conserve the fishery resource.[70] Price lists were distributed to canneries that set minimum prices for smaller sized shrimp, prices that exceeded the market price offered in other states. When the general market price for small shrimp fell below the union's price, the canneries shut down and the fishermen stayed at the dock. Thus smaller shrimp were allowed to mature and grow, providing for larger catches of higher priced large shrimp later in the season and increasing union members' profits.

There was a major effort, especially in Mississippi, to keep out-of-state fishermen from netting shrimp in the local waters. Shrimpers in major ports throughout the state were required by the union to sell their shrimp at the union's price, and union members could not sell to any canneries or packers not approved by the union. Processors who did not accept the union price could be cut out of the market effectively. Most important, shrimpers who were not union members were not allowed to sell shrimp at the dock; canneries were told whom to exclude. A fisherman in Mississippi either belonged to the union or had no local cannery that would buy from him; nor did out-of-state shrimpers have any place in Mississippi to sell their catch.

Union price fixing thus had a dual purpose in Mississippi; first, it raised the annual wages of union members by proportionately decreasing the profits of canneries; second, it served as a conservation measure to protect the state's shrimp stock.[71] In effect, the unions instigated their own open and closed fishing seasons, a strategy that resulted in higher wages because their fixed prices for large shrimp were below the market price for the same shrimp in other states.

In 1934 the Mississippi unions lobbied the state legislature for legal minimum-size counts for shrimp.[72] The legislature complied; no shrimp could be netted in Mississippi waters that weighed less than 40 per pound. The unions then attempted to enforce the new state law: they instructed union peelers not to work with undersized shrimp; they fined members who netted illegal shrimp; and they required boat captains who fished for small shrimp to carry on board a contract that included the union price.[73]

Louisiana shrimpers soon noticed that Mississippi shrimp were larger than those in their waters during certain times of the year and began to converge on Mississippi fishing grounds. When they unloaded the shrimp on Mississippi docks, fights broke out and violence was commonplace; the unions resisted the intruders.[74]

E. Paul Durrenberger documents the continuation of the conflict between shrimpers' unions and the processors after World War II; during the war prices were generally high because of the shortage of meat.[75] Strikes in Biloxi, Mississippi, and in Mobile, Alabama, occurred on August 15 and 17, 1946, when the Gulf Coast Shrimpers and Oystermen's Association and the Mobile Bay Sea Food Union both rejected price offers for their shrimp. The strike in Mobile lasted only six days, the one in Biloxi until the end of the month. In both cases the packers finally agreed to the unions' initial price offering. The following season, the Alabama shrimpers' union and the packers quickly came to terms, but in Mississippi the conflict over price began again. In 1951 there were price disputes in Mississippi throughout the fishing season, from April to late September, and a work stoppage that lasted for four weeks.[76]

When the federal government finally interceded, it acted in the best interests of the canneries and processors. In a series of legal rulings in the early 1940s the federal courts found the fishermen's unions guilty of violating the Sherman Antitrust Act. As Ronald Johnson and Gary Libecap have stressed, "[Shrimpers'] unions were particularly active from the 1930s through the 1950s and [might] have developed effective regulatory policies [over the fishery] had their actions not been consistently opposed by the federal government as violations of the Sherman Act."[77]

The log of cases against unions in other American fisheries provided legal precedence in ruling against the shrimpers' unions in the Gulf of Mexico, for example, *Manaka v. Monterey Sardine Indus-*

tries (1941) and *Columbia River Packers v. Hinton* (1942).[78] In 1946 a federal grand jury indicted the International Fishermen and Allied Workers of America Local 36 in Los Angeles for price fixing, arguing that the fishermen were not workers but independent businessmen who conspired against the processors and dealers.[79]

Of particular interest is *Gulf Coast Shrimpers and Oystermen's Association v. U.S.*[80] A lawsuit was brought against the Mississippi shrimpers' union in 1955; fishermen were accused of fixing the prices of shrimp and other seafood in direct violation of the Sherman Act. The court found the union guilty, a decision that upheld the 1940s' judgments against the union. The Fifth Circuit Court of Appeals specifically ruled that the shrimpers were independent entrepreneurs, not workers or laborers, and therefore could not organize to form a legitimate union.[81] It also declared that minimum size limits set by the union for the purpose of conservation (and which resulted in higher prices and wages) were not allowable.[82] Thus the shrimpers' unions in Mississippi, Louisiana, Alabama, and Florida were effectively crushed by the federal court decision and the series of decisions that had preceded it.

In Texas there were hard times even though no unions had been formed.[83] Shrimpers received low wages for their work.[84] The authors of the government report duly noted that shrimpers faced a difficult life in Texas as commercial fishermen but did not define the conditions as requiring any recommendations or solutions. Instead they suggested the need for "the formation of a strong organization of the industry," including the development of an industrywide accounting system, investigation into improving the handling of shrimp by canneries, a new system of catch statistics, and national advertising of the product. The report also raised concern over the depletion of the shrimp fishery: "We should look forward to the time when fishing becomes so intensive that the shrimp may be unable to withstand the drain."[85]

State government also took a renewed interest in the depletion of fish and wildlife stocks, creating the Texas Fish, Game, and Oyster Commission in 1920. But the new commission first focused on its most important commercial fishery, oysters. Not until the 1960s, when the commission became the Texas Parks and Wildlife Department, did it oversee a diversity of state-owned wildlife and fishery resources.

The bay shrimp fishery did not fall under the domain of the commission but remained under the jurisdiction of the Texas state legislature. Against the wishes and the interests of Texas bay shrimpers, the legislature set size limits in the early 1920s; by 1927 the limit was five inches.[86] Size limits generally hurt bay shrimpers in Texas because they discouraged the netting of smaller shrimp in the bays, the kind that the fishermen were most likely to catch in their nets. The Texas bay shrimpers had no unions, and without them, the fishermen were not adept at lobbying in their own best interests.

Texas bay shrimpers finally called for repeal of the size limits in 1947 and 1948, urging the Texas Fish, Game, and Oyster Commission to bring pressure on the legislature to repeal the existing law. The problem with low prices for small shrimp had been recognized by bay shrimpers, along with the canners and packers, since 1925.[87] Yet the fishermen were able to organize and to create pressure on the state legislature only when out-of-state shrimpers appeared ready to move into Texas. The competition from Gulf shrimpers motivated the bay shrimpers to unionize, but their attempts soon failed because shrimpers' unions in other Gulf states had been declared illegal by the courts.

Offshore commercial shrimpers from Louisiana, Florida, Alabama, and Mississippi relocated to the Texas coast immediately after the end of World War II.[88] The Gulf fishermen brought with them their oceangoing trawlers, larger boats built to net shrimp in deep waters. Gulf shrimpers gradually moved their base of operations to several key Texas ports that had convenient and direct access to Gulf waters, including Galveston, Aransas Pass, Port Isabel, and Port Brownsville.

For many years commercial shrimp fishermen had known of the abundance of brown shrimp directly off the Texas coast; white shrimp were less common, more difficult to find at sea. Shrimpers from other Gulf states came looking for new, and profitable, supplies. The war years had resulted in a scarcity of domestic meat, and market demand for brown shrimp, formerly disdained by many coastal residents and by Americans in general, rose when meat was in short supply. Out-of-state Gulf shrimpers were also motivated to keep their boats fishing year-round; they had invested much greater capital in comparison to the bay boat owners, and the longer the larger boats sat idle at the dock, the more money they lost.

By the postwar years the historical problems of spoilage and transportation to markets largely had been resolved. The oceangoing trawlers left the docks with reserves of ice in the hold that would last two to three weeks or more, depending on the weather. The ice was cheap, produced at the dock by machines that could manufacture as much ice as the fleets required. Thick sheets of ice, weighing more than 100 pounds each, were loaded on board by hand or, increasingly, by conveyor belts that led directly into the hold. Freshly netted brown shrimp were quickly deheaded by the crew on the rear work deck to delay spoilage, washed, and stored below between layers of ice. When the Gulf trawlers returned to the dock, they might bring with them 20,000 pounds or more of brown shrimp netted in a single trip. The shrimp were offloaded onto conveyor belts, sorted by hand or by machine, and iced once more before being shipped by refrigerated trucks either directly to wholesalers and retailers or to the processor, who was quickly replacing the canner.

Gulf state shrimpers were well aware of the untapped fisheries in Mexico's Bay of Campeche and of abundant shrimp fisheries farther south. More than a handful of fisherman had worked on Mexican shrimp boats or in neighboring Central American countries where they had seen the huge reserves of shrimp firsthand. It was not unusual for American commercial fishermen to work in Mexican, or Central or Latin American, fisheries in the 1940s, either directly for American corporations or for nationally owned fleets or fish cooperatives.[89] Various stories were exchanged among American shrimpers recounting the virtually unlimited supplies of shrimp along the east coast of Mexico and points south.

Shrimpers from Florida and Alabama traveled the greatest distance to reach the brown shrimp fishery off the Texas coast and the shrimp farther south. Only the most experienced were willing to cut across the Gulf of Mexico, saving hundreds of nautical miles but risking the high seas. Although offshore shrimpers often had considerable experience handling their fishing vessels in a variety of weather, they were first and foremost commercial fishermen, more rarely professional seamen with experience on the open sea. In order to reach the new fishing grounds, most shrimpers from Gulf states were forced to follow the curve of the Gulf shore, a trip of five to ten days, before they finally reached Galveston. From there, those who sought the plentiful shrimp in the Bay of Campeche had to motor 400 additional miles just to reach the mouth of the Rio Grande; after several more days of travel, they finally reached the Mexican shrimp fishery. The long trip was made profitable only by the large catches of shrimp that were almost guaranteed.

Commercial offshore shrimpers in Louisiana faced a much shorter trip to the brown shrimp fisheries, but some of them, too, saw the logistical sense of moving their operations to Texas ports. If their trawlers were based in ports in central or south Texas, they could continue to fish the traditional grounds in the Gulf of Mexico and still follow the migration of brown shrimp south to the Bay of Campeche and beyond. The southern ports of Texas initially had been considered much less desirable because the motoring distances from them to traditional fishing grounds in the northern part of the Gulf were far less cost-effective.

The movement of men, boats, and supporting operations and facilities was not a uniform or well-organized exodus. Most important, the families of shrimpers did not pack up their belongings, sell their houses, and move to Texas. There were false starts: some shrimpers and their families moved back to their home state after several months in Texas. Thirty years after the commercial shrimpers made the move, some still remembered the difficulties they had faced.[90] There were Florida shrimpers who first moved to Alabama, unwilling to move farther from their original homes. Others moved to

Texas, then decided to return to Florida. Still others first moved their operations to Mexico, then back to the Texas Gulf Coast. Frank Voltaggio, for example, moved his fishing business from New Orleans to Tampico, Mexico, in the late 1940s. He relocated to Brownsville, Texas, in the early 1950s, establishing one of the first shrimp processing plants in the area. By the early 1960s Voltaggio's Valley Frozen Seafoods operated the largest fleet, thirty-five Gulf boats, based in Port Brownsville and Port Isabel.[91] Other Louisiana shrimpers, many of whom were Cajun, moved to Texas ports and were discriminated against; some Cajuns soon moved back home to the bayous, sick of Texas and Texans.

The larger Gulf boats drew more water than the bay boats and required deeper channels and harbors, more dock space to unload the catch, larger fuel depots and ice houses, welding services, and net shops that could meet their specific fishing needs. The bay fleet did not need deep-water passes and channels through the barrier islands to reach the Gulf, but the new Gulf boats sought Texas ports with quick access to it. There were only a handful: Galveston, Aransas Pass, Port Isabel, and Port Brownsville. A few others such as Freeport had access but were at some motoring distance to the Gulf; at Corpus Christi and Port Lavaca, for instance, trawlers had to travel fifteen miles or more before reaching open waters.

The Texas fishing communities of the 1940s differed little from those in the 1920s even though the coastal population had changed significantly in the northernmost counties, primarily as a result of the growth of Houston and Galveston and, to a lesser degree, of Port Arthur. The total population of the coastal counties had grown to more than 1 million, 17 percent of the state's residents; however, just three of the eighteen counties, Harris, Jefferson, and Galveston, accounted for more than 70 percent of the coastal population. Significant increases from 1920 to 1940 also occurred in Nueces County because of the growth of Corpus Christi. Aside from these four counties, a majority of Texas fishing communities witnessed limited population growth during these years.[92]

The geographic isolation of the coastal communities continued to influence their social and cultural isolation.[93] The Texas towns had more in common with fishing communities in Louisiana, Mississippi, Alabama, and Florida than they did with the landlocked communities not more than ten miles inland. The towns were relatively homogeneous with regard to the dominance of the white population, and small town elites were firmly ensconced. As a result, change came slowly, if at all, to the communities that stretched from Sabine Pass to the Mexican border.

For the small Texas ports, the arrival of the Gulf shrimp industry was arguably the most important economic, social, and political event of the twentieth century, its overall impact superseding the discovery of offshore oil or the tourist industry that now dots the coastline. But as soon as the first Gulf out-of-state shrimpers moved to northern and central Texas, they met with stiff resistance, fueled by Texas bay shrimpers who saw them as competitors.[94]

The Gulf trawlers were capable of netting tens of thousands of pounds more shrimp than the tiny boat boats, and the local fishermen feared that the newcomers would quickly drain the waters close to shore. Many of the big boats also could trawl in the deeper bays, channels, and rivers, the traditional fishing grounds of the bay shrimpers. The Texans reasoned that since they did not fish in the waters of Louisiana, Mississippi, Alabama, or Florida, then the out-of-state fishermen should not be permitted to bring their trawlers to Texas waters. And the bay fishermen believed that substantial increases in shrimp offloaded by Gulf boats at Texas ports would lower prices for their own catch. They saw the new Gulf fishermen as unfair and unwanted competition.

A good portion of the newcomers were Cajuns from Louisiana, and Texas bay shrimpers, mostly white, viewed them as foreigners, people who spoke a language other than English, who practiced customs incomprehensible to them. Ultimately, from the bay shrimpers' viewpoint, fishing jobs were at stake. Racism, however, played a strong supporting role, making it far easier to dispute the intentions of the new fishermen, easier to treat them as second-class citizens because they spoke English with a strong accent, ate strange foods, and practiced strange customs.

Open hostility and overt violence toward the Gulf shrimpers was counterbalanced by the potential for a gigantic boost to the local economies of the communities. The Gulf shrimping industry brought with it an influx of jobs and dollars even as it threatened to destroy the local fishing industry. In Port Isabel, for example, workers were hired to build new docks to serve the larger Gulf boats, fish houses were constructed, and new workers hired. The town received an infusion of outside capital that it had never seen before as new dollars went to many different types of employers and workers.[95] And in Aransas Pass it was estimated that by 1952 one of every seven residents in the community was directly tied to Gulf shrimping; the $5 million Gulf industry there had a tremendous impact on the local economy.[96]

The Texas bay shrimpers fought long and hard against the Gulf competitors. Resistance took many forms, both in the courts and in supralegal activities. Gulf shrimpers and their families were initially ostracized by many residents in the Texas towns.[97] The newcomers were excluded from positions of responsibility and denied elective offices in the communities; their relative affluence did not immediately buy them social status or political influence. In Aransas Pass some shrimpers and their families were threatened with violence, and many of the Cajuns who first settled there were virtually driven out of town. A small group of Cajuns moved their boating operations farther south to Port Isabel, where they were somewhat better received; nevertheless, years later they were still outspoken about how long it took to gain acceptance in Port Isabel.[98]

Some Texas shrimpers had seen the handwriting on the wall and had begun lobbying against Gulf shrimpers as early as 1941 by convincing the legislature to enact certain gear restrictions on Gulf trawlers. The new laws limited the size of nets and regulated the size of net mesh on Gulf vessels, putting bay boats in a more competitive position even though the trawlers still netted more shrimp.

Texas bay shrimpers, now more motivated to organize than ever before, took their grievances to their elected representatives and in 1947 convinced the legislature that it should adopt a statute requir: ing a large fishing license fee from nonresidents. And two years later a statute was passed that limited the number of shrimping licenses available to non-Texas fishermen.[99] To the surprise of the bay shrimpers, the Gulf shrimpers appealed, and the statutes were thrown out by the Texas Supreme Court. Bay shrimpers then appealed the state court's decision, only to fail before the U.S. Supreme Court in subsequent decisions.[100]

The court battles and the community hostility and violence in the long run did not advance the bay shrimpers' position. The Gulf industry hired its own lawyers to fight for its best interests. The relatively huge economic impact of the industry in the small Texas fishing communities could not be denied; eventually even the Cajuns were accepted, if grudgingly, into positions of social and political influence. The Gulf shrimp industry lobbied long and hard before the state legislature, spending more time, effort, and dollars than the bay shrimpers could muster. Within less than a decade, the bay shrimpers' worst fears were realized and embodied in the Texas Shrimp Conservation Act of 1959.

The political influence of the Texas Gulf shrimp industry grew as its economic importance to coastal Texas became increasingly apparent. Though it took more than a decade in some coastal communities for the new shrimp boat owners to be fully accepted, in the legislature dollars spoke louder than state boundaries or ethnic origins. The owners of Texas Gulf boats formed the statewide Texas Shrimp Association (TSA), with local chapters in the Texas ports where the trawlers docked; it employed a full-time director who soon began lobbying the legislators on behalf of the Gulf boat owners.

The TSA persuaded the lawmakers to pass the Texas Shrimp Conservation Act, a piece of legislation that boldly represented the interests of the Gulf industry. The act was clear in its intent: for the first time legislation limited the months in which bay shrimpers could fish for white shrimp in Texas bays; further, it prohibited them from netting brown shrimp in the bays. Under the new act, bay shrimpers were allowed to fish for white shrimp, their traditional catch, only from August 15 through December 15. Gulf waters were closed to bay fishermen from June 1 to July 15, a time during which bay fishermen sometimes ventured out of the bays to net brown shrimp that had attained a larger size and thus a higher market value. Creating the law in effect allowed the brown shrimp to migrate farther offshore where the bay shrimpers did not usually venture. Bay shrimpers were also mandated by the act to maintain daily catch quotas of no more than 250 pounds. Catch quotas had never before been mandated on shrimp in Texas; the new laws did not set any quotas for the Gulf shrimpers.

Bay shrimpers were caught completely off guard by the act. They had no formal statewide organization, no paid lobbyists in Austin to argue for them.[101] Nevertheless, they were able to rally their forces, eventually convincing the legislature to allow them a limited spring season for brown shrimp, from May 15 to July 15. They also persuaded the lawmakers to raise the daily catch quota from 250 to 300 pounds.[102] Although the concessions were significant, the Shrimp Conservation Act was a major defeat for bay shrimpers that reflected their political powerlessness and the newcomers' strength.

The new laws were clear, but their enforcement was altogether another matter. The Texas Fish, Game, and Oyster Commission did not have the field agents to patrol the hundreds of miles of bays, inlets, rivers, beach fronts and fish house docks where bay fishermen practiced their trade. Even if they had possessed a larger number of agents, it was questionable whether in practice the new laws could be enforced adequately, especially once the bay shrimpers contrived their illegal fishing strategies.

Bay shrimpers ignored the new laws, which grew increasingly restrictive until the 1990s, because they judged them unfair, favoring the Gulf fishermen. Indeed, they were unfair for bay shrimpers; they had been designed to be so by the TSA. Still, the laws remained largely unenforceable, and attempts to carry them out were limited to gestures more symbolic than real. Bay shrimpers developed a strong disrespect for the laws governing their shrimp fishery and for the agents who attempted to enforce them. Over time, many bay shrimpers forgot that it was the TSA that had lobbied for the laws and instead often blamed the TPWD's agents for the catch quotas, closed seasons, and other restrictions.

❖ ❖ ❖

PROBLEMS IN THE GULF INDUSTRY

The Texas Gulf industry had its way with the bay shrimpers even if the new laws and the others that followed were not always enforced. There were more than enough shrimp for both groups; Gulf shrimpers prospered as they took advantage of the new Gulf of Mexico fisheries on both sides of the Rio Grande, following the shrimp as they migrated into Mexican waters and returning with boats laden with their catch.

Despite the efforts of the TSA, the trawler owners still had problems, chiefly the sporadic gun battles with the Mexican navy over the rights of American boats to fish in the Bay of Campeche. At least one American shrimper was shot and killed in the late 1950s.[103]

Although Gulf boat owners prospered, their crews did not share equally in the profits, and the Texas Gulf shrimp industry soon faced a series of labor strikes. Even though shrimp crews were banned from forming unions, a spontaneous strike, one of several, broke out in Port Brownsville and Port Isabel in 1960 over low wages and working conditions on the trawlers. Several boat owners were accused of treating their crews unfairly. Thirteen years later another strike occurred at these same ports over similar issues.[104] In both cases, crews attempted to form unions but were defeated because of previous court decisions that prohibited their organizing.[105]

The Tax Reform Act of 1976 added to the problems that Gulf crews faced; it contained provisions that reduced the wages of Gulf crew members. Shrimpers along the Gulf were redefined as self-employed workers and were exempt from employment withholding tax. Thus crews could no longer rely on unemployment insurance or on the Federal Old-Age, Survivors' and Disability Insurance Program.[106] Crews threatened to strike in several ports over the Tax Reform Act but no strikes occurred. Texas Gulf boat owners looked upon the 1950s, 1960s, and early 1970s as the golden years of shrimping, when production was high and prices were relatively stable. Gulf shrimp crews, including large numbers of Mexican workers in ports in South Texas, were kept in their place; labor was cheap.

The 1980 census recorded the total population of Texas' coastal counties as almost 4 million, more than a quarter of the total population of the state. Over 60 percent lived in Houston or the surrounding area. Jefferson County boasted 250,000 residents, Galveston County an additional 200,000. With the exceptions of Cameron County, where the majority of the population was in Brownsville and Harlingen, and Brazoria County, the home of numerous smaller communities such as Lake Jackson and Freeport, the population in the coastal communities remained basically unchanged. Except for the small towns near Galveston and the Houston metroplex and those in South Texas close to Corpus Christi, most of the fishing communities were still physically, and to a large degree, politically and socially isolated from major urban areas. The handful of communities that harbored large numbers of Gulf boats gradually incorporated the newcomers into their political and social lives; most of the other small towns and villages remained centered on the bay fishery.

Although there was real and lasting hatred between bay and Gulf shrimpers generated by competition over the same fishery, hostilities between the two industries after the Shrimp Conservation Act were surprisingly limited. Bay shrimpers broke the laws and dodged the marine agents while beyond the barrier islands, the Gulf shrimpers plied their trade. The bay shrimpers made a decent living, the Gulf boat owners got rich, and their crews barely got by.

The relative calm came to an abrupt end in the late 1970s for a variety of reasons as government policy and regulations played an even more crucial role in the lives of bay shrimpers. First, Vietnamese refugees were resettled in many Texas coastal communities, fishing villages already suffering from a nationwide economic recession. Prices for shrimp plummeted, forced down by imported shrimp. Local and national groups of recreational and sports fishermen, environmentalists, and other activists appeared on the scene accusing the shrimpers of destroying the environment. Each of the interest groups demanded that the Texas legislature and the federal government regulate the shrimpers, both bay and Gulf fishermen. The new players lobbied for the passage of a series of laws in the 1980s and 1990s that greatly restricted commercial shrimping in Texas. In response, the Gulf shrimp industry sought alliances among the interest groups and renewed its efforts to legislate the bay shrimpers out of existence. Bay shrimpers fought the settlement of the Vietnamese, dodged the new laws and the marine agents as never before, and tried to organize. And for the first time, they lost money when they worked, their established ways of living threatened by national and global forces beyond their influence or their understanding.

6

New Alliances and the Vilification of

Shrimpers

❖ ❖ ❖ ❖ ❖ ❖ ❖ ❖ ❖ ❖ ❖ ❖ ❖ ❖ ❖

Fishing is the oldest industry there is known. It's the only thing that's not government controlled. One thing that's free. Now they want to control that.
—*Interview with a bay shrimper*

The Magnuson Fishery Conservation and Management Act redefined the role of federal and state government in the shrimp fishery. Passed by Congress in 1976, the act embodied a broad national concern for the marine environment by establishing a 200-mile zone that extended offshore from the U.S. coast or its barrier islands. Within this Exclusive Economic Zone (EEZ), access to fisheries by foreign fleets was severely restricted. More important for Texas bay shrimpers, the Magnuson Act established eight Regional Fishery Management Councils that were mandated to create Fishery Management Plans (FMPs) for each fishery within their regions.

The councils were made up of political appointees and were subject to lobbying. Members included owners of fish house operations and fleets, upper level managers of local, state, and federal programs and agencies associated with some aspect of the marine environment, and a handful of academics.[1] The councils functioned as quasi-governmental agencies, developing policy that was then turned into regulations governing the use of marine resources, including fisheries.

The council that oversaw the Texas Gulf of Mexico shrimp fishery, located in Tampa, Florida, was given broad but unexplored powers under the Magnuson Act. Although it did not directly control Texas bay shrimpers, it did manage the Gulf fishery. Each time the Texas Gulf trawler owners, represented by the Texas Shrimp Association, felt their industry being jeopardized by council rulings or other federal policies or legislation, they pressured their competitors in the bays.

The most influential regulations passed by the council, or which came under its auspices, were the rules covering Turtle Excluder Devices (TEDs); TEDs regulations were soon followed by those governing By-Catch Excluder Devices (BEDs). Texas state legislation concerning spotted trout and red drum fish had preceded TEDs and BEDs, revealing new statewide and national interest groups that had formed in the late 1970s and 1980s. The Lacey Act, amended by the U.S. Congress, also indicated the nature of the new alliances, outlining the political strategies that the new interest groups would use against commercial shrimpers.

Legislation at the state level from 1975 to 1983 culminated in House Bill (HB) 1000, which banned any sale in the state of red drum and spotted trout netted in Texas. A commercial fishing industry that traced its roots to the time of Stephen F. Austin thus was legally destroyed. Eight hundred thousand Texas recreational fishermen and their associations, among the most powerful of which was the Gulf Coast Conservation Association (GCCA), had defeated 1,700 Texas commercial fishermen.[2]

Conflict between saltwater recreational fishermen and commercial fishermen was not new in Texas. As early as 1893 recreational fishermen had complained that the commercial harvest of red drum sorely depleted the stock: "Believing that the fish are caught in greater quantities than their natural fecundity can make good, there is a desire on the part of many persons, especially those interested in developing the sporting fisheries of Texas, to restrict in some way the use of seines." The argument fell on deaf ears, however, because of the economic importance of the fishery to local fishing communities. If the commercial fishery were closed it "would throw entirely out of employment over 350 men, removing from the coast towns a monthly revenue of more than $12,000, and taking from the market a cheap and wholesome article of food."[3]

Texas recreational anglers before the turn of the century failed in their efforts, but ninety years later other sports fishermen made the same charges against commercial fishermen, blaming them for disastrous declines in fish populations along the coast. The GCCA was at the forefront of this systematic assault. Formed in Houston in 1977, its first chairman was Walter W. Fondren III, a wealthy oilman and the heir to the Exxon fortune.[4] The original founders, influential Texans and weekend saltwater anglers, worked aggressively to acquire new members, and by 1979 the GCCA had 1,000 members. More important, it had the financial resources and political savvy of a core group of Texans dedicated to their cause. The GCCA, in contrast to other grass-roots groups that formed about the same time, immediately hired lobbyists to argue for its interests in the state legislature.

The GCCA hired Richard "Dickie" Ingram as executive director and Jack Gullahorn and his Austin law firm of Roan and Gullahorn to lobby in its behalf in 1980, meanwhile continuing its membership drive so that it had grown to 23,000 by 1981.[5] With its numbers increasing rapidly, a national organization soon formed, composed of other sports fishermen who were alarmed about declining sports fish populations.[6]

Legislation to regulate the commercial fishermen who harvested red drum and trout preceded the formation of the GCCA, but it was the association, through its strong political influence, that was instrumental in the move to outlaw commercial netting of red drum and spotted trout in Texas. From its inception the GCCA proved to be an exceptionally influential lobby, far more so than other groups or organizations with which it sided, including local coastal sports fishing groups and freshwater fishing clubs. Among the other organizations that provided funds and political expertise to work for the ban on commercial drum and trout fishing and sales was the Texas Shrimpers Association.[7] The TSA hoped to court favor with the GCCA and, possibly, to weaken any future attempts to develop legislation to regulate the Gulf shrimping industry.

Chief among the GCCA's political tactics was its manipulation of the Texas Parks and Wildlife Department, The TPWD had grown rapidly since its inception, and by the 1980s it employed more than 2,600 people and was financed by an annual budget exceeding $100 million; about 500 of its employees were game wardens.[8] It was overseen by the Texas Parks and Wildlife Commission (TPWC), which was composed of nine members appointed by the governor and approved by the state senate. The executive director of the TPWD was appointed by the nine commissioners and served at their will.

During the time spanning the passage of the legislative bills that curtailed and then eliminated commercial drum and trout fishing, several members serving on the TPWC were recreational fishermen and strong advocates for the GCCA, but the apparent conflict of interest did not restrain them from making their decisions. Regulation of the red drum and spotted trout industry came under the Coastal Fisheries branch of the TPWD, and, with the direction of the TPWC, the department provided capable technical assistance to the GCCA throughout the legislative process that eventually ended in 1983 with HB1000.

Before 1975 the TPWD had advocated a statewide policy that freshwater fishing in the interior of Texas was the domain of sports fishermen while saltwater fishing along the coast and in the Gulf of Mexico was under the purview of the commercial fishermen and their industries.[9] Then the TPWD changed its traditional position, siding with the recreational fishermen and the GCCA although there were serious concerns within the TPWD over the issue. Among the people who thought the TPWD shifted its position too quickly were several of the department's own fishery scientists, who questioned the assumption, advocated by the GCCA and other groups, that commercial fishermen were directly responsible for the decline in fishery stock. They also questioned whether in fact there actually had been a decline in the fish population.[10]

The GCCA, however, exercising its influence over the TPWC, "captured" the Texas State Department of Wildlife and used the agency's status and resources to advance its own causes.[11] The GCCA and its fellow organizations and groups based their arguments on two assertions. First, they claimed that fish populations were declining, according to the best scientific studies available, and they blamed the commercial fishermen for the decline.

Yet they ignored a study by the National Marine Fisheries Services (NMFS) that directly contradicted findings in studies by the TPWD. The NMFS study showed that saltwater fishermen were themselves responsible for catching more than 75 percent of all red drum.[12] In effect the NMFS study argued that any declines of fishery stock were attributable at least in part to increase in the number of saltwater fishermen and in the number of fish they netted, legally and illegally. The credibility of the TPWD's own research efforts was called into question when their scientists admitted that their data had a 50 percent margin of error.[13] Such a large margin clearly invalidated any conclusions based on the statistical analysis in the study, voiding any claims premised on the research.

The GCCA's second assertion, a position it also argued most effectively before the state legislature, was that the sports fishing industry generated more than $800 million in the Texas economy; the drum and trout commercial fishing industry was shown to account for only $11.8 million.[14] The TPWD demonstrated that the unemployment rates in the eighteen Texas coastal counties would increase only by an insignificant .05 percent if the commercial fishing industry were shut down, raising unemployment from 5.39 to 5.44 percent.

The $800 million figure, a projected estimate based on a variety of economic assumptions, was extremely generous and certainly subject to serious questioning. The unemployment data were also quite misleading, purposely deemphasizing the real impact of lost jobs on local fishing communities and towns. For example, it was true that the impact on Harris County of fifty or more newly unemployed commercial fishermen would be negligible; the county is home to the Houston metroplex and fifty lost jobs would have been slight. But for communities such as Port Mansfield and other small coastal towns, the loss of fifty jobs would constitute a considerable impact on the local economy, a situation as true in the 1980s as it was in 1893 when recreational fishermen first claimed that commercial fishermen were destroying the fishery. Again, the TPWD was most generous in providing GCCA with the economic expertise it needed in order to make its case against Texas commercial drum and trout fishermen; an $800 million sports fishing industry carried a lot of weight in a state legislature concerned with a regional economy in serious decline.

One may also question the validity of an argument based on the assumption that one industry is ipso facto "better" than another because its economic impact is significantly larger, in this case $800 million compared to $11.8 million. Such logic neglects the importance of any social or cultural values that in fact may be held in greater esteem than a difference in dollars. The worth to a society of a particular industry and its labor force can be far greater, or far less, than its economic value. Indeed this same logic was soon advocated by the GCCA and other groups in the case of sea turtles, which have little real economic value. But the TEDs issue skirted the economic-impact argument and centered on a different value: environmental concerns.

Although the series of legislative bills covered a period of eight years, more than enough time to secure funding for and to conduct a comprehensive study on the drum and trout fishery, no new scientific studies on the fishermen and their fishery were ever initiated. In truth little was known about these fishermen, who they were, how they fished, or how their commercial fishing directly affected fishery stocks, if at all. The impact of a transition to other kinds of employment was never seriously considered: How would the fishermen and their families adapt to other industries? Was alternative employment available? What about the impact on local economies? The TPWD estimated that 1,000 fishermen and their families would be affected by closing down the industry, but again, the figure was only a best guess in lieu of any legitimate study of the labor force and the industry.[15]

Instead of facts generated by legitimate research, collected and documented in an objective fashion, the TPWD relied on a wide variety of nonscientific sources, including the conjectures and accounts of recreational fishermen themselves, most of whom of course had little if any real knowledge about commercial fishermen. Given the lack of research about the drum and trout industry, its public image was easily manipulated. A picture was created in the popular media that portrayed the fishermen and their families as poachers, marginal members of their communities who destroyed fish for a few dollars a day. And once they were defined as marginal, legislating their work out of existence was not difficult. Dick J. Reavis has documented the process by which every commercial fisherman became an environmental rapist and every saltwater angler became a conservationist.[16]

The recreational fishermen, led by the GCCA, designed a legislative strategy based on their facts and figures to eliminate the commercial fishermen over a period of a few years. Each new piece of legislation further constrained the drum and trout fishermen, who were faced with the same dilemma that bay shrimpers later faced; either they had to abide by each new law and lose income, if not their business, or they had to break the law and risk fines. Each new regulation that was passed increased their risk; fines and penalties grew stiffer as new restrictions on the fishing season, size limits, and size of catch were enacted.

In this no-win situation, drum and trout fishermen chose to break the new laws; by doing so they could at least stay in business and hope for better times. And so these commercial fishermen, fishing in the same ways as they had before the turn of the century, became under the new regulations "the new rustlers," their wives and children partners in crime.[17]

The GCCA effectively lobbied for more TPWD agents because the commercial fishermen continued their illegal fishing. The increasing number of fines, dutifully recorded by the media, demonstrated that the fishermen were lawbreakers; additional TPWD agents were required to catch the fishermen. Texas newspapers continued to record the number of gill nets and the pounds of confiscated fish taken from the outlaw fishermen, furthering their image as common criminals, as poachers.

With the continued technical assistance of the TPWD, the GCCA argued that the only solution to the increase in lawlessness was a total ban on all commercial fishing of drum and trout. House Bill 1000, the last of ten pieces of legislation spanning eight years, thus ended the commercial harvesting and selling of red drum and spotted trout in Texas and established penalties for anyone who broke the newest set of laws.

The bill passed despite the facts, ignored by the GCCA, the TPWD, and the state legislature. For example, from 1976 to 1979 Texas commercial fishermen, including shrimpers, were cited 1,137 times for illegal fishing practices; the fact that was not publicized was that sports fishermen in Texas were cited 1,600 times, typically for fishing without a license or for exceeding the legal limit. Even though TPWD field agents focused on catching the commercial fishermen, they also caught sports fishermen who were breaking the laws. The image of saltwater angler as archconservationist was, at best, misleading; fishery stocks were being depleted by the very people who were supposedly protecting them for the public.

The GCCA, with the help of the TPWD, the Texas Shrimpers Association, and the other groups and organizations that supported HB1000, had created a masterful obfuscation, but their forces did not go unopposed. At least fourteen groups and organizations formed to fight the long series of measures culminating in HB1000, among them the Political Involvement of Seafood Concerned Enterprises (PISCES).

PISCES garnered strong support from bay shrimpers, their loosely knit local groups, and other coastal organizations, but they were grossly underfunded, compared to the GCCA, and politically ill-prepared to do battle with their powerful rival. The GCCA spent about $2 million in its lobbying efforts, in contrast to the $300,000 mustered by the fourteen other groups, including PISCES. Moreover, the GCCA donated an additional $1.2 million to the TPWD for a red drum hatchery. They also provided the TPWD with a helicopter to monitor illegal commercial netting in the Laguna Madre. In a move further hurting the cause of the commercial fishermen, Perry Bass, a Fort Worth billionaire and one of the original board members of the GCCA, was appointed as chair of the Texas Parks and Wildlife Commission.[18]

PISCES received as much support form the bay shrimpers as they could provide, as evidenced from the survey conducted in 1987 and 1988. Twenty percent of the bay shrimpers interviewed belonged to PISCES, an additional 25 percent to other groups of bay fishermen who supported the commercial finfishermen, including a group called Concerned Shrimpers. The bay shrimpers interviewed reported in general that they had supported the red drum and spotted trout fishermen because they were worried that the same political tactics might be employed against them in the near future.

Several of the more outspoken and articulate shrimpers said they believed that the Texas Shrimpers Association sided with the GCCA in order to build up favors with that powerful organization. The bay shrimpers reasoned that after the drum and trout fishermen had been removed legally from their means of livelihood, the bay shrimpers' longtime enemy, the Gulf shrimpers, with the help of the GCCA, would go after them. They also believed that when the time came the TPWD would side with the GCCA and the TSA. They were correct.

The decisions made by the TPWD and the TPWC about the series of laws ending in HB1000 were typical.[19] In the same year that HB1000 shut down the red drum and trout fishermen, the TPWD made a series of decisions about closing the bays to recreational fishermen so that fishery stocks could recover from a serious freeze. Then, reacting to the strong political arm of the sports fishermen, the department quickly reversed its rulings.[20] Other positions of the TPWD from 1983 to the 1990s reflected several different emerging strategies that would reappear frequently in the coming years and that confirmed the department's close working relationship with the GCCA and the TPWC. When the use of a fishery was called into question, economic-impact arguments were deployed as the primary ammunition; the industry that could be shown to bring the largest number of dollars to the Texas economy was by definition superior to any competition. When studies completed outside the TPWD contradicted the department's aims, they were ignored or discredited, as were fishery scientists within the TPWD who disagreed with the prevailing view. Commercial fishermen were depicted as poachers, vilified in order to justify legislation that would destroy their industry. New scientific studies of the fisheries would have clarified the difference between fact and social myth, but new studies were never commissioned.

❖ ❖ ❖

THE LACEY ACT

For more than thirty years Texas shrimpers had relied upon the Mexican fishery as a valuable source of their annual income, up to 25 percent or more for many Gulf shrimpers who fished out of Port Isabel and Port Brownsville.[21] Out-of-state shrimpers who had moved to Texas ports in the 1940s and early 1950s to be closer to the abundant fishery in the Bay of Campeche were not disappointed by their decision to relocate. Texas Gulf shrimpers continued to harvest large amounts of shrimp from Mexican waters, frequently joined by shrimpers from other Gulf states.

The shrimpers did not break American law when they fished in the Bay of Campeche, but they did break Mexican law. Texas shrimpers frequently fished within the territorial waters of Mexico, which extended ten miles from land. Over the course of four decades, however, business practices and customs that superseded the law were established between American shrimpers and officials of the Mexican government, including the Mexican navy and local officials in Tampico and Veracruz.[22] Occasionally the arrangements broke down; one result was the "shrimp wars" of the 1950s. Yet generally the best interests of the involved parties were served.

Texas trawler captains fished openly in Mexican waters; the boat owners knew where their boats were going when they crossed the Rio Grande. Most of the time the American boats were ignored by the Mexican navy, charged with protecting Mexican sovereignty; there was more than enough for all the fishermen. The Mexican shrimp fishing fleet, organized in cooperatives, found no shortage of shrimp to catch, but lacking modern fishing gear and electronic navigation and location systems, they were able to harvest only a small portion.

When the Mexican navy did choose to catch Texas shrimpers within the ten-mile limit, or wherever they found the Yankee boats, they boarded the trawlers with the expectation that they would be generously bribed. Texas captains knew that money sufficient enough to satisfy the Mexican gunboat captain and his crew was expected. If no bribe was offered, if it was not large enough, or if some other offense was given, intentional or not, then the Texas trawler was escorted to the docks in Tampico or Veracruz.

At the Mexican docks the wise Texas captain, or the American owner who immediately came to his aid, would offer a bribe if he expected to avoid additional problems. The amount had to be larger than the bribe a gunboat captain would accept because once land was reached, more Mexican officials had to be satisfied. Texas shrimpers knew that once they docked in Tampico or Veracruz not only would they lose their catch to the Mexicans but much of their gear also would be taken. It was far better to offer a substantial bribe at sea than it was to lose time and more money at a Mexican dock.

The custom of the bribe, or *mordida*, a familiar practice among Texans and Mexicans in a variety of business transactions, worked quite well for many years. Both parties benefited. Although Texas shrimpers lost money to the gunboat captains, they kept their catch, gear, boats, and crew; bribes were a part of the overhead. Shrimp were so easy to net that Texans knew losses could be recouped in a few short weeks or less. The Mexicans were satisfied because they benefited, albeit indirectly, from their shrimp fishery; they did not have the boats to net all the shrimp, so the bribes served as an unofficial fee charged to foreign commercial fishermen.

Then the government in Mexico City announced in 1978 that fishing in Mexican waters would henceforth be by permit only. Mexico, mimicking the actions of other nations, including the United States, extended Mexican territorial waters from 10 to 200 miles. The new plan called for a three-year phase-out of the official permits, a limited number of which were sold to Americans, after which no American fishing boats would be allowed to fish in Mexican waters.

Although the permits were phased out by the Mexican government in 1981, many Texans continued to shrimp south of the Rio Grande. The Mexican navy, to save face before the federal officials in Mexico City, enforced the new laws but without much enthusiasm. Bribes were still given and accepted although much more cautiously; Mexican law had changed, but the familiar method of doing business remained the same.

In early fall 1981, just as the phase-out of permits to fish in Mexican waters reached its last stage, the Lacey Act was amended by the U.S. Congress. The original act had been conceived by Congressman John F. Lacey in 1900 to control the interstate traffic of wildlife products since in the 1930s the laws of one state did not extend to others. The act was then amended in 1935 to include the trafficking between the United States and foreign countries of wildlife obtained illegally. Similarly, the Black Bass Act of 1926 had been created to protect black bass taken, purchased, sold, or possessed in violation of state law; it was eventually expanded to cover all species of fish and in 1969, like the Lacey Act, was amended to include foreign commerce.

Supporters of the newly amended Lacey Act successfully argued that the trade in wildlife products had become so profitable that the penalties under the original laws, including the Black Bass Act, were no longer meaningful. The maximum penalties under the original Lacey Act had called for a fine of $10,000 and one year in jail; a perpetrator was guilty of a misdemeanor. Anyone breaking the laws set forth in the Black Bass Act could be fined up to $200, sentenced to three months in prison, or both.[23]

The 1981 Lacey Act amendments made it illegal to import exotic wildlife products and to net shrimp in Mexican waters.[24] Under the new laws, Texas shrimpers who fished in Mexico were breaking not only Mexican law when they fished in the Bay of Campeche but also American law when they recrossed the Rio Grande. Civil penalties for engaging in conduct prohibited by the new laws included a fine of up to $10,000. The criminal penalty, now a felony, called for a fine up to $20,000, imprisonment for up to five years, or both. If a shrimper was convicted, he might forfeit his fishing gear and his vessel.[25]

Texas Gulf shrimpers did not take the Lacey Act seriously. They believed it would never be enforced, that the laws were preposterous; shrimp were not an endangered species. The fishermen were genuinely shocked when the U.S. Coast Guard, under the direction of the National Marine Fisheries Service, began arresting them as they recrossed the border from Mexico. Before the arrests, Gulf shrimpers had looked upon the U.S. Coast Guard as their allies on the ocean who helped them when their engines failed, when a man was lost overboard, or when a shrimper was seriously injured and required medical attention.

The Coast Guard stopped more than thirty boats in spring, summer, and fall 1982, arrested their captains and crews, and confiscated their catches. There followed an attempt by the Gulf industry and the NMFS to reach some sort of working compromise, but talks broke down and the arrests and confiscations continued into 1983.

The dispute was picked up by the national media as hostilities peaked. Anonymous threats were made against the crews of Coast Guard cutters and their land-based families. There was at least one bomb scare at Port Brownsville, and someone tried to burn down the NMFS office there.

The shrimpers were at first shocked by the enforcement of the new act and then angered by the rough treatment from overly zealous law enforcers. Law officers at the state and federal levels knew little about shrimpers and their industry, and a few individuals were particularly inexperienced. The Coast Guard crewmen, for example, had orders to board the Texas trawlers using the same procedures required in seizing elusive and often heavily armed drug-smuggling vessels on the open seas. When the Coast Guard boarded the shrimp trawlers, they carried loaded M-16s; they were covered by 50-caliber deck machine guns on the cutters. Jumping down from the deck of the cutters to the smaller Gulf trawlers, the Coast Guard yelled at the shrimpers to follow their orders exactly or risk being shot. The crew were held at gunpoint while their boats were searched. At the dock, the Gulf shrimpers were handcuffed in front of their waiting and anxious families before they were taken away to jail in Brownsville. The shrimpers thought it was unjust that they were being treated as if they were hardened criminals when they had done nothing more than catch some Mexican shrimp. It was senseless to them that such a tremendous effort of time, money, and men was expended to protect shrimp that the Mexicans themselves cared little about.[26]

Additional Coast Guard vessels were brought in from the Gulf area along with NMFS agents from other parts of the country. American military jets were routinely used to identify and track shrimpers in Mexican waters. More than fifty shrimp crews and trawlers were seized in 1983.

Gulf shrimpers in Port Isabel and Port Brownsville took the federal government to court; Judge Filemon Vela, the first Hispanic ever appointed to a federal district, heard the case in Brownsville. The courtroom was packed with angry shrimpers listening to legal arguments they did not understand. Outside in the halls shrimpers argued about the injustice of the Lacey Act, often having to be silenced by courthouse security. A handful of lawyers from Washington,

D.C., flew in to represent the interests of the government; the shrimpers hired local lawyers whose knowledge of federal law unfortunately was less than impressive.

Judge Vela, politically astute, controlled the packed courtroom crowd well, at times amused by the Washington lawyers, at times chiding them, suddenly and sternly overruling their motions, which prompted the shrimpers to nudge each other and smile. The judge treated the local attorneys with a respect that came from living in the same community with them and their clients. To a man the shrimpers believed that justice would prevail, and they left the courtroom at the end of the four days more than confident that their side had won. Six months later Judge Vela ruled in favor of the federal government, upholding the Lacey Act and the government's right to seize Texas trawlers returning from Mexico.

In fall 1983 the U.S. Department of Commerce (USDC) ruled that confiscated trawlers could keep their catches although fines and felony charges were still to be collected for those individuals who had broken the law. Some years later the majority of fines against Gulf shrimpers either were completely forgiven by the federal government or owners were allowed to pay a much smaller penalty; felony charges against captains and crews were also dropped. The details of the negotiations were kept quiet by the parties involved, and the media gave them little attention.

In the 1990s many Texas shrimp boats once again fished the Bay of Campeche during the winter months, filling their holds with thousands of pounds of Mexican shrimp before returning to American waters. If caught by the Mexican navy, the *mordida* would be expected as usual. Upon recrossing the Rio Grande the Gulf captains were not detained by the Coast Guard, nor were fishing activities monitored by jet flyovers; the men, vessels, and airplanes deployed at great expense in 1982 and 1983 had long since been reassigned.

The short-term impact of the Lacey Act amendments on the Texas Gulf shrimpers was significant. The men who fished out of southern coastal ports, from Aransas Pass south to Port Isabel and Port Brownsville, roughly two-thirds of the Texas Gulf shrimp fleet, were hit hard. At the worst extreme, some owners went out of business; a majority were adversely affected until national attention ebbed and the political interests that had forced the National Marine Fisheries Service to take action in the first place finally subsided.

Texas Gulf shrimpers lost more than money and in some cases their businesses; the Lacey Act, the methods by which it was enforced, and the ways in which the media covered the actions served to vilify Texas Gulf shrimpers. The public never understood what was involved because the real issues were never examined by the media or in a court of law. The legitimacy of the Lacey Act itself and the political reasons for amending it to include shrimp were never reasonably questioned. Shrimp, whether American or Mexican, were certainly not endangered. But in the minds of an increasingly environmentally sensitive public, the fishermen were assumed to be in the wrong; the media never attempted to suggest otherwise. Dirty, unshaven shrimpers were shown at the dock after they had been arrested and handcuffed. The visual image did not call into question why they had been arrested or the justification for it; the shrimpers looked dangerous. On the other hand, violence against the Coast Guard, or the threat of violence, was carefully documented by the media. Shrimpers were portrayed as criminals, and it was assumed that the Coast Guard, local law enforcement officials, the NMFS, and the Washington lawyers were doing their jobs. Media analyses never indicated that the shrimpers were hounded for doing theirs; indeed they were portrayed both in print and on television and radio as the men who stole shrimp from poor Mexicans.

The participating federal agencies, in an attempt to justify their involvement in the enforcement of the Lacey Act, released statistics to show that enforcing the act had helped stem the flow of illegal immigrants from Mexico to the United States. Records released by the NMFS revealed that undocumented Mexican workers were arrested on Texas boats as they recrossed the Rio Grande after fishing in Mexico; the NMFS was alleging that shrimp boats routinely ferried over undocumented workers. The undocumented workers arrested by the Coast Guard on Texas boats were in fact full-time crew members, headers and rigmen. The captains had hired them in south Texas ports, transported them south, and were returning to their home ports when stopped by the Coast Guard. When the shrimp season was over, the Mexican fishermen returned to Mexico as they had been doing for the last forty years. By raising the issue of illegal immigration, NMFS hoped to rationalize its decision to enforce the new act and to deflect any possible criticism.

Meanwhile, the Lacey Act was not the only problem confronting the Texas Gulf shrimpers. Vessel insurance premiums continued to rise, along with a growing number of lawsuits brought against Gulf boat owners, at least some of which juries judged to be without merit.[27] Prices continued to decline, and investors, often with little knowledge of the fishing industry, bought new boats. The result was more Gulf boats, each sharing a smaller piece of the pie.[28]

❖ ❖ ❖

SENATE BILL 609

The Texas Gulf trawler owners felt that their industry was under siege from all sides; they had to fight back to regain their lost profits or soon they could not afford to keep shrimping in the Gulf of Mexico. The path of least resistance led straight to the door of the bay shrimpers; if they could be legislated out of business, there would be more shrimp to catch in the Gulf and profits could be reclaimed.

Bay shrimpers had some measure of protection from the state legislature against the legal attacks of Gulf shrimpers. The latter could lobby harder and longer than the underfinanced bay shrimpers, however, and most often they got their own way. Although the lobbying was done far from the public's eye, the lawmakers' actions were scrutinized as they made their way through the house and senate. In the light of this public forum the bay shrimpers, although always the underdog, at least had a chance to make their case, to advocate their own best interests.

The solution for the Gulf shrimpers was to remove the bay

shrimp industry from the protection of the state legislature with its public forum and to place it instead under the aegis of the Texas Parks and Wildlife Department. The TPWD was subject to the political pressure generated by the TSA's lobbyists and the industry's political friends who served on the Texas Parks and Wildlife Commission, but it did not have to answer directly to the public or its elected representatives.

Senate Bill (SB) 609, which came to a vote in late April 1985, removed regulation of the bay shrimp industry from the legislature to the Texas Parks and Wildlife Department. Proponents of SB 609 argued that it would put regulation of the industry into the hands of the state agency that had the most expertise in handling the fishing industry. Senator Hector Uribe, a strong advocate of the Gulf shrimpers from his district, which included Port Brownsville and Port Isabel, believed that if the Texas legislature continued to regulate the bay shrimpers it would lead to "the destruction of the shrimping industry." He presumably meant that Gulf shrimpers would not make as much money as they had in the 1950s, 1960s, and 1970s if the new bill did not pass. Further, the senator argued, "the Legislature meets every two years. We should set broad policy and delegate the day-to-day administration to the agencies that have been created to develop expertise in these areas." Uribe also noted that "bay shrimpers have *notoriously* captured and harvested shrimp before they reach full maturation."[29]

A filibuster was organized by a handful of legislators who sympathized with the bay shrimpers. Carlos Truan, senator from Corpus Christi, held the floor of the state senate for twenty hours in hopes of persuading legislators to change their minds. Truan talked on as almost all the senators left the chamber, his only audience a large group of bay shrimpers who had docked their boats and driven to Austin to lend their support. The filibuster was a half-hearted effort at best, bound to fail. The bay shrimpers were poorly organized; even if they had developed a more coherent political strategy, they lacked the money necessary to negate the effects of TSA lobbying. But they did not give up. They knew that once the

TPWD held sway over their industry, their modest political influence would disappear in an instant; decisions would be made within the confines of a large state bureaucracy by politically appointed officials already biased against their interests. Senator Truan summed up the situation when he observed in his speech that "it's just the big guy wanting to kick the little guy out of business. That's all."[30]

Although the filibuster was a futile effort, it was entertaining in a way that is perhaps unique to Texas politics; a fight broke out in the antechamber between two legislators after the filibuster. Sen. Carl Parker, Port Arthur, and Sen. Craig Washington, Houston, were among the few who supported Truan's filibuster. When it ended, Parker chided Sen. Hugh Parmer, Fort Worth: "Senator, you keep it up, and you ain't never going to be elected to the same job twice." Parmer replied to Parker, who had recently been reindicted by a grand jury on drug and pornography charges, "Senator, I may not be reelected, but I ain't been indicted either." While the senators fought, the frustrated bay shrimpers yelled their encouragement.[31]

Still, the bay shrimpers did what they could to forestall the inevitable. It took four years to transfer the regulatory powers from the state legislature to the TPWD, the bay shrimpers fighting it every inch of the way. As a further vexation, the TPWD game wardens, long the recipients of shrimpers' taunts and hatreds, now enforced policies and laws that regulated the bay shrimpers, rules engineered by their own agency. It was just a matter of time, the bay shrimpers believed, until they were driven out of business legally. Then TEDs appeared and made their situation even worse.

❖ ❖ ❖

TEDS

The otter trawl, other innovations such as the motorized winch, and the more contemporary electronic location and navigation devices increased the efficiency and production of the fishing effort.

Shrimping remained relatively unchanged, however, a crude and unsophisticated affair accomplished in the same basic ways practiced by the first commercial shrimpers. Long and cumbersome nets were laid out, dropped over the side, and dragged along the bay or ocean bottom. By nature a hit-or-miss business, the fishermen's money, time, and safety were bet on the hope that when the net was winched over the side it would be brimming full of shrimp.

Shrimpers could never count on boarding nets filled with shrimp, but given the way they fished, they knew that there would always be plenty of trash fish, one of the curses of their work. Shrimpers netted a wide diversity of marine life; anything that was on or near the bay, river, or ocean bottom might end up in their nets along with sand, mud, weeds, and rocks. Of the hundreds of pounds of catch boarded, only a small percentage of the total was marketable shrimp. It has been estimated that bay shrimpers netted trash fish at the ratio of four to one, Gulf shrimpers twelve to one.[32] The hundreds of pounds of trash, after culling, were swept over the side. The majority of it was already dead, having been first towed in the nets for several hours, winched up quickly from the depths of the water, and finally dropped on the work deck to lie in a pile for an hour or more until the shrimper had time to sort through it. Only the jellyfish were likely to be unharmed, along with the crabs that scurried about the deck.

Shrimpers, in effect, swept the bays and Gulf with large vacuum cleaners that did not discriminate among marine species; fishing technology was primitive, far from perfect. One drag in the Gulf produced more than 1,000 pounds of trash fish, from which might be culled less than 100 pounds of shrimp. In the bay the shrimper's single net might bring in 300 or 400 pounds of trash fish for every 100 pounds of shrimp; I saw at times less than 15 pounds of shrimp culled from drags on the bays.

Only the largest and smallest of fish and other marine life escaped the shrimpers' nets. The smallest were sucked into the mouth of the net, but they escaped through the mesh. Even very small fish, less than two to three inches in length, were often captured in shrimp nets. Larger fish avoided the slow-moving trawlers although again I saw small sharks and other species of fish up to two and a half feet in length brought up in the nets.

Texas shrimpers, both in the bays and in the Gulf, also caught turtles in their fishing nets.[33] Since a single trawl might last up to several hours, sea turtles, which must regularly take in oxygen, could become enmeshed in the nets and drown before the net was retrieved. Turtles that were still alive when brought up could be revived and returned to the sea with minimal harm done to them.

Texas shrimpers did not fish for sea turtles, but they sometimes landed one in their nets because of the inexact nature of their fishing gear. From a shrimper's perspective, a sea turtle was just another minor problem that was quickly remedied, a bit of the work that had to be done.[34] Sea turtles were part of a small but viable industry in Texas that thrived from the 1860s through the turn of the century.[35] Beefpackers on Aransas Bay, for example, began canning sea turtles in one-to-two-pound tin cans in 1869. Texas fishermen during certain seasons used special nets to trap the turtles in Aransas and Matagorda bays and at the lower end of the Laguna Madre along the Mexican border. Captured turtles were kept alive in cages underneath Texas docks until they were shipped to market or canned. In 1890 a turtle cannery at Fulton processed 243,000 pounds of turtlemeat. The majority of sea turtles was shipped by steamer to Galveston to the fresh-fish markets although some reached markets as far away as New York City.[36] To the south, along the Mexican coastline, sea turtles are still caught and used for food although protected by Mexican law.[37]

Concern for declining populations of sea turtles in Texas waters was first expressed more than 100 years ago. According to a government report, "Green turtle are gradually becoming less abundant on the coasts of Texas, yet on account of the increasing demand for them the annual catch is probably increasing."[38] The Endangered Species Act (ESA) was passed by the U.S. Congress in 1973, and it fell to the National Marine Fisheries Service, under the umbrella of the Department of Commerce, to provide supervision and protection for the turtles, covered by the ESA.

In 1978 the six species of sea turtles found in American waters, which included those in Texas bays and the Gulf of Mexico, were listed as endangered or threatened. Scientists agreed that one species, the Kemp's ridleys, was closest to extinction; these nested only on one stretch of beach near Rancho Nuevo, Mexico. From the late 1940s to the 1980s the population of nesting Kemp's ridleys females on this isolated beach, now closely guarded by Mexican soldiers, declined from 40,000 to 500.[39]

Although sound scientific evidence confirmed that sea turtles, especially Kemp's ridleys, were declining in population (the rates of decline were in dispute), the major argument was over the exact cause and thus the solution to the problem. Environmental groups increasingly blamed shrimpers, including Texas bay and Gulf shrimpers. The fishermen, in their own defense, pointed out that though they did catch turtles in the course of their work, it was a rare occurrence, and they argued that factors outside the industry were the major killers of sea turtles.[40]

Scientists and technicians for the NMFS began developing a turtle-excluder device for shrimpers in the late 1960s. In its most basic form, the TED was a small box, approximately two to three feet square, constructed with a metal frame and fishnet sides and a trap door on top; it was placed at the mouth of the trawl net. A turtle that became entrapped in a shrimper's long net in theory would immediately find its way through the trap door to escape unharmed. Shrimp, on the other hand, were pulled farther into the net where they remained until the bag of the net was released on the work deck of the boat.

Contraptions not unlike the NMFS's TED commonly had been used by Texas shrimpers for many years, devices called "cannonball shooters" or just plain "shooters." Cannonball jellyfish fouled a shrimper's nets and were an inconvenience that soaked up his time and energy; the jellyfish made the catch so heavy that it strained the winch, lines, and nets and made it difficult to board the bag. During certain seasons shrimpers might run into cannonballs by the hundreds. Texas shrimpers, along with other shrimpers along the Gulf of Mexico and the eastern seaboard, developed cannonball shooters to separate the jellyfish from their catch. The major problem with the shooters was that they required "fine tuning"; through trial and error and the fisherman's expertise, the device had to be set into the nets in such a way that it did its job and at the same time it did not allow shrimp to escape.

The NMFS continued to work on problems with the TED and by 1983 began urging Texas shrimpers to use the device voluntarily. Shrimpers were reluctant to use it, even though for many years some had used shooters, partly because they did not like the idea of the federal government intervening in their fishing practices and partly because they questioned the effectiveness of the NMFS's TED. They worried that it would cost them shrimp in their nets, and according to the stories they had heard, it required extra time and expertise to fish it properly.

Texas shrimpers also questioned their role in the decline of the sea turtle populations. The majority of the fishermen did not deny that they sometimes netted sea turtles and that they could die despite attempts to revive and return them to the water. Many shrimpers believed, however, that there were other more important causes for turtle deaths, including residential and industrial development along the coast, offshore oil industry pollution in the Gulf, and the practices of Mexican shrimpers.

The decline of sea turtles became an important symbol for the problems of America's coast and ocean waters in the mid-1980s, rivaling the public's concern with the decline of the American bald eagle or of the whooping cranes that wintered in the coastal marshes near Rockport. The major causes of sea turtle mortality were circumscribed, and environmental groups targeted shrimpers as their major enemy, as killers of defenseless marine creatures, even though scientists agreed that other factors might be blamed. It was much easier for environmental groups to vilify Texas shrimpers, who constituted a small industry, than it was for them to face off against a phalanx of Exxon or Shell lawyers representing offshore oil interests or petrochemical plants along the Texas coast.

The NMFS acted with bureaucratic caution; in light of the shrimpers' recent response to the Lacey Act, the agency had learned a hard lesson about the fishermen and therefore was reluctant to attempt to force them to accept anything they did not want, especially a mandate for TEDs. Shrimpers on many different occasions had been outspoken in their resistance to TEDs, and there was good reason to believe they would not accept the devices even if stiff penalties and fines were to be imposed. Environmental groups grew more impatient with each bureaucratic delay, at last threatening to sue the NMFS and other government agencies if the use of TEDs was not enforced soon.

Negotiating sessions sponsored by the federal government were held between the environmentalists and the shrimpers in late 1986.[41] During this time reenactment of the Endangered Species Act was initiated in Congress, and Gulf shrimpers lobbied their legislators to delay its passage and thus the enforcement of TEDs. More environmental groups joined the fight as fishermen filed countersuits. Various levels of government entered the fray, including the state of Louisiana.[42]

By summer 1989 shrimpers had exhausted their legal alternatives. On July 22, during the middle of the summer shrimping season, Gulf shrimpers were told they had to begin using TEDs immediately. The announcement led to spontaneous protests in Houston, Aransas Pass, Corpus Christi, and other Gulf ports. Some bay shrimpers joined the Gulf shrimpers who lined their boats side by side across boat channels, blocking traffic going in and out of Texas' major ports. In a few cases, shrimpers violently resisted the U.S. Coast Guard.[43]

Enforcement of TEDs was suspended in deference to the protests, this time by Department of Commerce secretary Robert Mossbacher, but in early September TEDs were again mandated.[44] Finally, the environmentalists believed, the turtles were going to be protected from the shrimpers' nets. This time Gulf shrimpers did not protest by blocking ship channels, but apparently they were not using TEDs during the grace period they were allowed; enforcement by the Coast Guard then became the issue.[45]

Meanwhile environmental groups moved to force the use of TEDs in coastal bays and other waters regulated by the state, presenting the same arguments against bay shrimpers as they had against Gulf shrimpers. Bay shrimpers responded that they should be exempt from TEDs because the turtles rarely entered the bays and estuarine systems; the environmentalists rejected their defense.[46] When John Sununu, chief of staff who had been holding up the decision on TEDs at the White House, was fired, it appeared that the environmentalists would get their way.

The Mexican government then announced that it would enforce the use of TEDs on its shrimp boats, and the environmentalists cheered the news. The Mexican government, however, lacked the means and the will to carry out such enforcement so that despite governmental intention Mexican shrimp boats do not use TEDs.[47]

The strategies that national and state environmental groups employed to support the use of TEDs on shrimp trawlers were remarkably similar to those used to pass the red drum and spotted trout bill. The only difference was that the groups had grown in membership. The Concerned Shrimpers Association numbered about 2,300 members throughout the Gulf of Mexico in 1991, but the Gulf Coast Conservation Association had grown to 17,000 in Texas alone.[48] Coalitions between recreational and sports fishermen were formed, providing the environmentalists with an influx of cash to support their legal and lobbying strategies.

The GCCA in Texas quickly sided with the other national and state environmental and sports fishing groups but not before dropping the Texas Shrimpers Association, its former friend and ally. For some time the TSA had studiously courted the support of the GCCA, siding with it against the commercial drum and trout fishermen and the bay shrimpers, but now the TSA found itself out in the cold. Bay shrimpers, for their part, understood that if laws could be passed that forced Gulf shrimpers to fish with TEDs, then they most certainly would be next. For the first time in many years, Texas bay and Gulf shrimpers joined forces against a common enemy.

The media were effectively used as a part of the vilification of commercial fishermen. Images of the rare turtles, seagoing reptiles, were sentimentalized far beyond their dictionary definition: "Any of various reptiles of the order Chelonia, having horny, toothless jaws and the body enclosed in a bony or leathery shell into which the head, limbs, and tail can be withdrawn in most species."[49] The long-time work of one particular advocate, the Turtle Lady, on South Padre Island and her network of dedicated helpers and supporters exemplified the strategy of anthropomorphism. They made every effort to bring out the "human" qualities of these coldblooded animals teetering on the brink of extinction, dressing them in doll clothes and handling them as if they were babies. For example, one turtle was dressed in doll clothes, including a hat, and held closely by a woman as if it were a newborn infant; with the help of a human hand, it waved greetings with its front flipper to visitors. A variety of costumes were custom-made by volunteers for the turtles, who modeled them on the "Johnny Carson Show."

The anthropomorphism of sea turtles was effectively marketed to children, one of the primary target audiences. In the popular children's magazine *Ranger Rick*, a publication of the National Wildlife Federation, shrimpers were portrayed as fishermen who appreciated the benefits of TEDs and were looking forward to the next piece of technology that they would be mandated to use. In one story three cheerful animals, each one a different species, rowed out to inspect a shrimp trawler. The raccoon, Ranger Rick, wore a hat monogrammed with his initials; his friend Scarlett Fox wore a handsome red scarf around his neck. Along with Sammy Squirrel, they watched shrimpers prepare to fish. Their sea gull friend, Gabby, told them, "I knew the nets would each have one of those funny TEDs inside. That stands for Turtle Excluder Device. . . . If the sea turtles get caught in the net, it lets them escape."[50] Sea turtles, as portrayed here and in previous stories in the magazine, were the equals of Ranger Rick and his cute friends; to net and kill sea turtles was to do the same to them, a moral truth of this children's tale. The shrimpers in the story were good human beings because they used TEDs. As harmless as it was on the level of pure entertainment, the story clearly contained values held by the National Wildlife Federation and was written not only to educate but also to persuade and convince its young readers.

Particularly significant in the advocacy of TEDs was the manipulation of scientific data. Its intentional misuse had been a successful strategy during the public debate leading to the demise of the redfish and trout fishing industry. Perhaps the most noteworthy example of manipulation took place under the aegis of the National Academy of Sciences (NAS). In an attempt to stall the enforcement of TEDS, Senator Howell T. Hefflin and other observers in 1988 argued that a legitimate and unbiased review of the scientific literature was required to determine turtle mortality and other related issues in question. At a cost of about $800,000, researchers under the direction of an eleven-member committee appointed by the NAS completed a supposedly state-of-the-art report on turtle mortality. It was delivered in May 1990.[51]

Prior to the report, scientists had estimated that about 11,000 turtles per year were killed by drowning in shrimp trawler nets. The new report, which did not undertake any new research efforts, increased the estimate by a factor of three to four times. The researchers argued that previous estimates had been in error and that between 33,000 to 44,000 turtles were killed annually. The committee of scientists, in effect, simply reviewed the existing scientific literature, found it misguided, and issued a new estimate. In so doing, they faulted previous research yet did not conduct any new research on which to base their findings. The report concluded that shrimpers were the largest single contributor to sea turtle mortality.[52] In advocating the necessity of TEDs, supporters clearly misused scientific data. For example, in public hearings involving Texas and Florida shrimpers seeking an injunction against the use of the devices, the government spokesmen readily manipulated data in charts, tables, and bar graphs on the catch efficiency of TEDs and on turtle mortalities; their conscious manipulation of data can best be described as voodoo statistics. Compilations of meaningless data were presented in the hope of persuading, perhaps even confusing, the judge who heard the case. The problem of contradictory data or of the reliability of data based on estimates of estimates was never seriously challenged.[53]

What was perhaps most remarkable was that new scientific studies were never requested or carried out, a failure that had also occurred in the cases of the Texas finfishermen and the bay shrimpers. New studies that clearly could have shown the impact of TEDs, if any, on production, efficiency, vessel safety and profits, and viability of the industry or of the mandate's social effects on family and community were never initiated. Instead, $800,000 was spent to fund an NAS report that reviewed the existing literature and reached startling new conclusions.

The impact of TEDs on the shrimpers was unequal. Bay shrimpers indeed had lent their support to Gulf shrimpers, first to stop and then to delay the use of TEDs. Yet despite the numerous stories and reports in the media to the contrary and despite their public complaints, bay shrimpers were minimally affected by having to use TEDs. They trawled with only one net, compared to the four hauled by Gulf shrimpers; using TEDs meant more work for them and probably less shrimp. But the devices were a bigger problem for the Gulf shrimpers than they were for the bay fishermen.

Data in the random sample survey conducted in 1987 and 1988, during the height of the TEDs debate, support the assertion. Bay shrimpers interviewed revealed attitudes toward TEDs different from the rhetoric often produced by their representative groups, voiced by individual shrimpers in public hearings, or discussed among other shrimpers at the dock. To the question "What is the most important problem of controversy bay shrimpers now face?" the most common response, 34.2 percent, was "existing laws" regulating the limits of pounds of shrimp that could be netted. The second problem named was harassment by Texas Parks and Wildlife Department agents, at 17.1 percent, and the third was "too many boats" fishing the bays, at 10.5 percent. TEDs ranked fifth in importance, at 7.8 percent, falling behind the 8.5 percent who said pollution was the biggest problem or controversy.

Bay shrimpers were much more familiar with TEDs than was popularly believed. A third of the sample, 33.1 percent, said they had used a TED to help them get rid of cannonball jellyfish. When asked, "Will you use a TED during the first year of implementation?" at a time when they were not required in the bay, almost a third, 29.9 percent, answered that they would. Finally, almost seven out of ten, 67.6 percent, answered that they would use TEDs when they were required by law to do so, compared to 16.9 percent who answered no; 15.6 percent said they were undecided.

Bay shrimpers sampled nevertheless were direct in their evaluation of TEDs. When asked, "What is your opinion of TEDs?" their answers included "I can't see where they would help us or the turtle"; "I don't like them. They'll end up hurting somebody"; "It's a bunch of bull. I maybe caught 8 to 9 turtles my whole life"; "It ain't worth a shit"; and "There is a lack of knowledge at NMFS to put this on the fishermen." Several bay shrimpers, at least in private, were staunch supporters of TEDs. One said, "They're good for what they were intended for," and another reflected a strong environmental concern when he responded that "I don't believe the turtles deserve to be extinct. I think they're a good idea."

The majority of bay shrimpers declared that they would use TEDs, but every shrimper interviewed agreed that the devices would have a negative impact on production; most disagreed only on the size of the impact. When asked, "How do you think TEDs will affect production?" their responses included, "You lose between 20 to 30 percent, if not more. You have to fine tune it everyday"; "don't take much to foul nets. Lots of lumber and plastic in the river"; "Maybe 10 percent"; "It'll cut down 25 percent to 30 percent of my income"; "[I'll] end up starving to death"; and "They're going to cut it down 50 percent to 75 percent."

Bay shrimpers might have been less biased against TEDs than perhaps expected because they thought, with good reason, that their use would or could not be enforced. There is also the distinct possibility that the shrimpers believed they could develop methods in their day-to-day fishing that would allow them to evade the law, much as they had done in the past with other restrictions.

The bay shrimpers' answers to the questions in the survey were also a measure of their political sophistication. A majority of the shrimpers interviewed, 51.3 percent, blamed the existing laws and the law enforcement agents for their major problems; only 1.3 percent pointed the finger directly at "environmental groups." The bay shrimpers also saw economic conditions as generally of no real importance. Only 1.3 percent of those interviewed blamed imports as the most important problem in the shrimping industry, ranking it at the very bottom of their concerns.

The major effect of TEDs upon bay shrimpers was indirect. Gulf shrimpers, however, who firmly believed that TEDs would cut their production and profits by significant amounts, began to think of ways in which they could recoup their losses. Although TEDs had momentarily brought bay and Gulf shrimpers together, the Gulf boat owners no longer had to contend with the Texas state legislature to rid themselves of their longtime competitors. They had a significant presence in the Texas Parks and Wildlife Commission that oversaw the bay industry. Eliminating the bay shrimp industry was the easiest approach for Gulf shrimpers in their attempts to reestablish their profits.

❖ ❖ ❖

BEDS

If TEDs were the major controversial issue of the 1980s, then BEDs, by-catch excluder devices, may be its replacement in the 1990s.[54] The same interest groups that advocated TEDs have jumped on the BEDs bandwagon, noting that the shrimpers' nets bring up a diversity of species of trash fish captured along with the shrimp.[55] The TEDs regulations meant that endangered sea turtles were less likely to die in shrimpers' nets but did nothing to address the question of trash fish.[56] Gulf shrimpers caught trash fish and shrimp at an estimated ratio of twelve to one, bay shrimpers four to one, and were accused of depleting a variety of fishery stocks to the point of extinction. Red snapper was the first species designated by scientists as in dangerous decline.[57] Reacting to published reports and lobbying by environmental groups, the Gulf of Mexico Management Council restricted commercial snapper fishermen to thirteen-inch size limits as early as 1984, about the same time as the TEDs controversy was heating up. By 1990 the council set the limits for commercial snapper fishermen at 3.1 million pounds and restricted sports anglers to a seven-bag limit per day. That same year, after declaring it would reduce the commercial quota further and also limit the number of fish a weekend fishermen could catch, the council held public hearings along the Texas coasts to gather additional comment.

The council proposed to close the shrimp fishery for three months, arguing that, based on the best scientific data available, shrimpers were major contributors to the decline in the snapper fishery. They were accused of netting millions of juvenile red snapper as trash fish that died before they could be returned to the sea.

By a slim margin, ten to seven, the council temporarily took a softer approach toward the commercial snapper industry, the sports fishermen, and the shrimpers and decided not to close the waters to shrimpers in 1990. That same year, amendments to the Magnuson Act forbade the council or any other governmental body from requiring shrimpers to use BEDs or from shortening their season in order to protect the red snapper population; the amendments were in effect until 1994. Sports fishermen, including the GCCA, and environmental groups pressured the federal government to mandate the use of BEDs, not just to protect red snapper but as a safeguard for all species that were perceived to be in danger. Lobbyists argued that shrimpers were depleting the stock of a large variety of species of fish to the point of extinction, even if unintentionally. In their own defense, shrimpers questioned the validity of the scientific data that has been collected. Although the studies that had been completed were not totally convincing, there was little to suggest that new studies would weaken the environmentalists' claims. Trash fish have always been a fact of life for shrimpers, an acceptable cost of fishing with otter trawls. The real danger for shrimpers, both in the bays and in the Gulf of Mexico, is that the technology of BEDs has not been adequately developed by scientists. Shrimpers fear they will soon have to fish not only with TEDs, which they claim has cut down on their production, but also with BEDs. In lieu of the development of BEDs that satisfactorily separate juvenile fish from shrimp, bay shrimpers could either have their season cut short or perhaps eliminated. No equivalent of the Turtle Lady and her entourage have yet appeared to defend redfish or the other species netted by shrimpers; turtles can be outfitted in clothes and their bodies manipulated to simulate human motion, but fish offer far fewer options for anthropomorphism. Stories about the benefits and wonders of BEDs have already appeared in the *Ranger Rick* series, however.[58]

Texas bay shrimpers have become marginal commercial fishermen, a half step away from losing their legal right to fish for shrimp in Texas bays, rivers, and estuaries. They and their industry in fact are in much worse shape than the Texas drum and trout fishermen were shortly before the demise of the finfish industry. All else failing, the drum and trout fishermen could appeal to the general sympathy of the people of Texas through the state legislature. In contrast, the bay industry must plead its case to the Shrimp Advisory Committee, which in turn reports directly to the Texas Parks and Wildlife Commission. Both the state agency and the committee are in the firm control of sports fishing interests and the Texas Shrimp Association.

An incident involving the Valley Sportsman Club can perhaps illustrate how the current system works. In 1991 club members, reversing their original position, attempted to reopen to bait and bay shrimpers an area near the Arroyo Colorado in south Texas. The waters there had recently been closed to commercial shrimpers and designated as a shrimp nursery through legislation supported by

A bait stand and shrimp boat at Sargent, Texas. Wells under the metal roof are filled with live bay shrimp used for bait by recreational fishermen.

sports fishermen. The members of the Valley Sportsman's Club who fished in the Arroyo Colorado and the surrounding area were forced by the new law to buy their bait shrimp from as far away as Port Isabel, twenty-five miles distant, because the handful of bait shops and shrimpers in the immediate area found it difficult to harvest enough shrimp in the surrounding waters that were not off limits to them. Commercial bay shrimpers could not provide the sports fishermen with a large enough supply of bait shrimp for the weekend anglers to catch drum, trout, flounder, and other species regulated by the TPWD, nor could the local bait stands keep up with the demand even though they trucked in shrimp and sold it at ten dollars a quart.

The club membership objected to this price and wanted to change the very law they had recently helped to pass. Ed Cooper, club spokesman, knew that the Shrimp Advisory Committee would lend him a sympathetic ear; he expected and got a full hearing before the members, the majority of whom were sports fishermen themselves. There was only one bay shrimper on the committee. After Cooper explained the club's plight, the committee moved to redesignate part of the shrimp nursery, thus allowing commercial shrimpers to harvest shrimp from the area and supply sports fishermen with bait for their fishing trips. There was considerable support for Mr. Cooper's request until Harper reminded the committee of a point that had been overlooked in their discussion: "This committee and Parks and Wildlife [are] mandated to manage by scientific data."[59]

Committee members then decided it was important to seek scientific expertise. A study had been done twenty-five years earlier, but its author, who was present at the meeting, admitted he was not sure if the data applied in the 1990s or if it would support the redesignation. More discussion followed. Several committee members who agreed with Cooper pointed out that if they gave him what he wanted, which they hoped to do, then other bay and bait shrimpers might ask for the same privilege, namely the opening up of waters that had just been declared closed to commercial shrimping. Some members worried that they would then be sued by bay and bait shrimpers.

Ed Cooper's request was finally voted on; it failed by a vote of five to four. Given the circumstances, it was indeed remarkable that Cooper almost won. The sympathy of the Shrimp Advisory Committee obviously lay with Cooper and his Valley Sportsman's Club. The committee, in fact, asked the representative from TPWD to check into their legal vulnerability and also to see if additional scientific studies could be found that might support Cooper's request. The advice of the solitary bay shrimper on the committee was ignored, his often sardonic criticism of Ed Cooper's proposal falling on deaf ears.

In the 1990s the laws protected the sports fishermen and their supporters, including the Gulf shrimpers. Indeed, Ed Cooper and the Valley Sportsman's Club, on the slightest whim, felt confident enough to attempt to amend the laws they had helped to craft. They were almost able to do it, failing by only one tiebreaking vote. Bay shrimpers have been reduced over a period of years to a labor force with little political clout and with no laws left to protect them. In the vital game of state and federal governmental policy and policies, Texas bay shrimpers have been big-time losers.

7

Women Who Shrimp the Bays

❖　❖　❖　❖　❖　❖　❖　❖　❖　❖　❖　❖　❖　❖

You need to know the Texas way: kids first, then husband, then yourself.
Honor and hard work.
—*Interview with a fisherman's wife who shrimps*

Women play an integral part in fishing families and communities.[1]
They raise the kids, keep house, and carry out support roles necessary for social survival. The girlfriends and wives of Texas bay
shrimpers often keep the family books, help process or sell the
catch, act as onshore liaisons and purchasing agents, and are aware
of political issues relevant to the industry. Their activities are shaped
by the constraints of the bay shrimping industry, which in some respects allows them more independence than they might find in
other kinds of work.[2]

Besides serving as "shore skippers," some Texas women also
fish the bays with the men or captain their own boats although female captains are the exception rather than the rule.[3] Women
shrimpers were at first overlooked; women on bay boats who were
helping men unload the shrimp at the dock, cleaning up, or culling

---

This chapter was researched and written by Andrea Fisher Maril,
Department of Sociology, University of Tulsa.

were assumed to be carrying out land-based support roles. The researchers, although well acquainted with shrimpers and their work, were neither looking for nor expecting to find female fishermen. At first they did not.

When the fishermen were asked if they had wives, girlfriends, sisters, or mothers who had shrimped the bays, a significant number responded early in the survey that they did, alerting the researchers to a pattern of behavior they did not expect to find. A subsample of female Texas bay shrimpers was subsequently collected.[4]

The women who fish the bays are identical to the men except for three important demographic characteristics.[5] First, the majority of women in the subsample are between the ages of forty to forty-nine, with smaller age clusters in the early twenties and mid-sixties. In comparison, male shrimpers are relatively evenly distributed throughout all age categories from twenty-five to sixty. The difference is probably related to childbearing since women are more likely to report working on boats either before children are born or after they graduate from high school.

The second difference suggested by the subsample is marital status. Almost all the women are married, compared to about three-quarters of the bay captains; the minority who are not married were cohabiting with fishermen. Almost without exception, the women come to shrimping via family ties, following husbands or boyfriends into the industry or working for their brothers or fathers. Women whose ties with fishing are severed, for example through death or divorce, are expected, according to the respondents, to quit fishing. The one female in the subsample who continues to fish with no tie to a man is ostracized by all the "decent" women for trying to act like a man.[6]

Third, the job histories of female shrimpers substantively differ from their male counterparts. The majority of women in the subsample were employed as pink-collar workers in traditional women's occupations, including waiting tables, secretarial jobs, and store clerking, for which they were paid minimum wage. After marriage to a shrimper, the women most often become involved in shore-based enterprises such as running bait stands or fishing camps, peddling fish, or running a cafe on the dock that caters to bay shrimpers.

❖ ❖ ❖

WORK PATTERNS AND GENDER EXPECTATIONS

When taken within the context of information gained in the in-depth interviews, the marital status of female shrimpers, their age clusters, and their job histories strongly suggest that women who shrimp the Texas bays have different long-term work patterns from those of male fishermen. The differences become most apparent early in their careers, then peak during childbearing years. Such differences result from both the cultural definitions of gender identity on and off the boat and the nature of the work that shrimpers do.[7]

Ties of kinship are essential. Male shrimpers learn how to shrimp from their fathers; nearly half the men in the survey fit this pattern. In contrast, only a minority of women who shrimp learned from their fathers or brothers. The majority of women shrimp because they married men who fish for a living or men who became full-time bay shrimpers.[8]

Women are often initially attracted to their work by the money. One woman says, "In four to five days we made more money on the boat than I could make working six days a week, eighteen hours a day. And I was my own boss." Another says, "Why not? You go out there and a man gets paid $10 an hour, a woman $2.50. [But] in shrimping either you've got [the catch] or you don't." Another simply concluded, "[The catch] smells like money to me."

Male and female couples who shrimp know how much it costs to hire a deckhand and realize that the savings can mean the difference between just getting by and doing well. When a woman replaces her husband's deckhand, however, she is simply "helping out," working for the good of the family. The women do not see their work as shrimpers as a personal sacrifice but as part of what a wife is expected to do.[9]

Another benefit to couples besides extra family earnings is a stronger marriage. Jeanette, for example, says that she and her husband have developed a mutual respect for each other's work on the bay. She is proud that Al, her husband, has a reputation as one of the best fishermen around. Al believes, in return, that Jeanette is a hard worker and a good deckhand. Another woman who shrimps says, "I enjoy it, mostly because I can be with my husband. Just when it's roughest and you're ready to quit, your husband comes back on the deck and gives you a hug, a kiss, and you know how much he appreciates what you are doing." Another put it more directly, bragging that "I'm the deckhand . . . all the deckhand George has ever had."

A larger margin of profit does not explain why many of the women shrimp, however; a majority in the subsample had been shrimping for more than a decade, several for twenty years or longer. They are shrimpers for reasons other than the money saved by not hiring a male deckhand. The data also suggest that women who work on shrimp boats cannot be explained simply by the recent increase of women in nontraditional occupations. Each of the women who shrimped for twenty years or longer was able to name three to five other women who were shrimping when she began.

Fishing the bays, regardless of reason or motivation, was not easy for the women, especially when they first began. "There should have been a book written," lamented Jeanette. "It was horrible at first. I swear, I cried, I prayed, I wouldn't let my kids breathe on the boat." She kept asking questions of her husband, but he was a novice, too. When he yelled at her about something she did on the boat, she ran next door to cry on her neighbor's shoulder. As Meg pointed out, "You have to get used to a whole new life: no schedules, few facilities, like many boats don't have toilets." The women who keep at it have some combinations of relatives in the business, a sympathetic friend, or an understanding husband.

Female shrimpers also have to learn about money management

when they start shrimping. One older woman says, "I learnt a long time ago that if the money came in good one week, you couldn't go to town and spend it. You needed to save it for later, for the boat."

Women who learn the ropes are able to see a side of shrimping that they did not know existed. "Man, I love this shit," one asserted. "Where else can you go dressed this way, any way you want, barefoot and all?" At the same time, the female shrimpers see firsthand how difficult it is to succeed. One female hand says, "More and more people can't handle it and are being forced out of shrimping."

Despite some reservations about certain aspects of their work, women who shrimp openly declare their enjoyment of fishing on the bay. As Rose says, "To me, going out on a boat is fun. I get away from it all. There's a lot more stress on land, people calling you, interrupting you." The sentiment is repeated in an interview with another female shrimper. When asked about her work on the bay, she says, "It's a totally different world. People that fish, they stay calm." Another women responds that "it's a job if you think of it as a job, but I enjoy doing it so much that it doesn't feel like a job."

A woman in her sixties, reflecting on her work as a shrimper, says, "I'm just about the oldest one that works now. I know my shrimp. I know my fish, too. I'm not ashamed of the work I do. I love it, love the water." Another respondent says, "It's nice and quiet and peaceful. You're away from the city, the hustle, the bustle. It's like nothing else in this world." When asked what she liked best about shrimping, a different respondent says, "The being together [with her husband] when it's calm and peaceful and pretty out there. Making a living together."

Although women who shrimp seem to enjoy fishing the bay, gender expectations to some degree define the nature of the work they do. "There's nothin' a woman can't do [on a boat] unless she doesn't want to," asserts Kate, who grew up in a fishing family, shrimped with her husband for years on both bay and Gulf boats, and also captained her own boat. Rose agrees with Kate although she is more cautious in her assessment. Rose says that most boat duties can be done by a woman if she uses her brains rather than her brawn. Marcelle, with forty years of experience as a shrimper and shrimper's wife, concurs that technology has helped to simplify shrimping for women although savvy plays an important part as well. "They used to laugh at me, like when I got down on my knees to cull," she chuckles as she sits in the recliner in her cozy living room surrounded by pictures of her grandchildren and two frolicking dachshunds. "Not any more. Now people admire a man and wife who work together."

For women with the self-assurance of Kate, Rose, and Marcelle, the work of fishing is determined, as it is for males, by position on the boat. The work, whether one is a deckhand or a captain, includes maintaining and operating the boat, working the winch, dragging and picking up the shrimp, cleaning and repairing the boat and nets, culling, icing, and other necessary tasks. "I even know how to take a cullin' iron to the captain's head if he pisses me off," one wife notes with confidence.

Yet most women in the subsample feel that a woman cannot do some work, most often because they lack the muscular strength they believe is required. "Women can't handle heavy work," says one female shrimper, several others echoing this assertion. Paradoxically the definition of heavy work varies from one woman to the next. Most women, for example, willingly carry the shrimp in baskets around the boat or help transport it from the boat to the fish house; the baskets may weigh forty pounds or more.

The meaning of "women's work" seems to stem from cultural definitions rather than from the actual physical capabilities of male and female shrimpers. Winch work for most of the women interviewed, for example, is not defined as women's work. Women believe they do not have the expertise to use the winch, to lower the net into the water for a drag, to pick up the net when it is filled with the catch, or any other aspects of its operation. Winch work, "men's work," is the job of the captain, and most often when women shrimp the captain is a man, the deckhand a woman.

The female shrimpers in the subsample are protecting their stereotypical gender identity by denying that what they are doing in their work as shrimpers in any way violates the traditional conceptions of women. Women who cross the boundaries generate hostility in others. The logical solution to the dilemma is for the female shrimpers to keep a low profile, to deny any expertise they have in working the winch, doing their best not to outshine their male counterparts. Several of the women in the subsample take this position to an extreme, saying that they cull the shrimp and cook on board the boat but do no other kinds of work. Two of the women are young, new to fishing, and say they strongly dislike it. One of them says, "I hate it. You're always moving, messing with that smelly old shrimp." The third woman who defines women's work in a narrow way has fished for twenty years and has a widely known reputation for seasickness in all kinds of weather. Her penchant for seasickness excuses her from having to fish the bay regularly; she is available to her husband as a deckhand but only if the weather is perfect.[10]

At the other end of the continuum of female shrimpers, women who do not feel constrained by gender expectations are most likely to have been, or are, captains. Darlene, who learned shrimping from her father, found the responsibilities of being a bay captain worrisome at first, with other male shrimpers just waiting to give her a hard time, but she soon became accustomed to the pressures. "I did a real good job for my Dad. Never did tear up a net." Kate, in contrast, did not begin shrimping until she was thirty-three. "But ain't I good!" she exclaims, proud of her abilities and accomplishments as a shrimper.

The satisfaction of the women with their work as shrimpers is closely tied to their contribution as income earners. Shrimping for one woman is "a way of keeping score with dollars. You've got to be smarter than the shrimp. It's man against nature." She likes both the companionship of her husband and the fact that they make good money from their work as fishermen. Still, the women, even the captains, see the downside of shrimping. Most mention that at the end of their workdays they feel worn out, exhausted from their labor on the water. But the majority believe that they and their husbands have few real options. One says, "You take someone who doesn't have a college education and what else can you do? Strong back, weak mind, that's what this kind of job takes." The women view their work as often boring, repetitive, and mention the toll of the long and unpredictable hours. They also acknowledge the growing financial problems of the shrimping industry, agreeing with their husbands that it used to be a better life than it is in the present; most blame the new fishing restrictions for the change.

Not surprisingly then, only a handful of the women want their children to follow in their footsteps. One respondent says that if one of her kids shows an inclination to shrimp, she will tell him, "Son, run the other direction. Don't. I wouldn't advise anyone to get into it." Her husband, sitting next to her, nods his agreement. Mildred, another respondent, comments on the growing problems with theft, the decline in catch, the competition from the bigger boats, and price fluctuations. One sixty-year-old woman who has fished for many years with her husband sums up her life as a shrimper, wife, and mother: "It's hard as hell on a woman, but to make a living with the laws they've got—if she don't help, I don't know how they can do it. . . . Shrimping, it'll take it all out of you."

Yet despite the physical challenges, the definitions of women's work and men's work, the uncertain future of shrimping, and other reservations, almost all the women say they would do it again if given the chance. Though many of their male counterparts would choose other kinds of work to do, the women in the subsample believe that in general they have benefited directly from fishing on the bay.

❖ ❖ ❖

WORK AND FAMILY

Successfully learning the ropes or learning to negotiate gender dilemmas does not guarantee women an uninterrupted career as a shrimper; job changes were a common experience for nearly all the women who shrimp. Shelley, for example, worked as a secretary in

Houston and then as a waitress before she began to fish. At the time of the interview she had shrimped with her husband for many years. She was very conscious of her appearance, especially her hands: "My mother says my hands look like shovels," she says.

Jody did not work outside the home for many years because she was too busy raising her four children. Her husband worked in the oil patch and was away for long periods of time. In 1968 they took their life savings and bought a bait stand near Port Lavaca, along with a boat. It was not long, however, before she left the bait stand to her mother, preferring to work with Billy when he went out to fish the bay.

Darlene, in contrast, was born into a fishing family; her father was a shrimper, and she remembers that her mother would often shrimp with him. Darlene learned to shrimp from her dad and later learned even more from her husband. Before long, she captained a thirty-seven-foot bay boat. In the off-season she helps her husband build boats and run the family bait camp; she worked at a convenience store at one time in order to make ends meet. Darlene has three children, one still in diapers.

Like that of many women who shrimp, Darlene's work history reflects the intersection of employment and childbearing demands. She worked on a bay boat through most of her first two pregnancies. She kept her first baby with her onboard in a swing she rigged; after her second child she was forced to confront a serious dilemma. She could not keep working on the boat and at the same time care for her children; boats and toddlers are a potentially lethal combination. Yet babysitters and daycare are expensive alternatives, and as children get older, school schedules do not mesh with the demands of fishing the bay.

Young mothers in the subsample often continued to fish while their children were small. One toddler can be managed on a boat, but two are nearly impossible although some women were able to do it, at least from time to time. Mildred says that she washed many a diaper in salt water, bottles too, because she fished with her husband for two to three days at one time. But the majority of women must choose caring for their children over working as shrimpers, staying home to tend kids and to take care of onshore fishing business. One female shrimper says, "Children are too important to trust to others," a sentiment echoed repeatedly in the subsample, often accompanied by a comment about the dangers of babysitters and daycare.

As their children grow older, female shrimpers return to fishing, often bringing their children with them during the summer months when they are not in school. According to the women, this practice creates family ties that last long into adulthood. "We all pull together yet," says Mildred, describing how she and her husband helped their grown children to buy their boats and, in turn, how their offspring chip in to buy them new appliances for their kitchen.

Only a few of the women manage to continue shrimping without interruptions. Kate, for instance, sees nothing wrong with hiring a babysitter when she goes fishing with her husband. Darlene cut back on her fishing when she had small children but nevertheless keeps at it and plans to return full-time when her children are older. Generally, the women who fish the most, even though they have children, seem to be the ones who derive the greatest satisfaction from working as shrimpers.[11]

Once the children leave home, women often return to shrimping. Shelley, for example, had kids early, staying home with them and tailoring her onshore work to fit their needs. As her children became more self-sufficient, she once again began going fishing with her husband although not on a full-time basis.

❖ ❖ ❖

THE SECOND SHIFT

When female shrimpers return to the dock after a day or night of fishing, their work is not over. These commercial fishermen are expected by their husbands and children to fulfill their traditional duties and responsibilities as wives and mothers. One woman who worked for more than thirty years as a shrimper summed it up: "I love it, love that water. But shrimper's wife, you lead a hell of a life."

Female shrimpers make do, especially during the fishing season. For most of them, the care and supervision of children always comes first: "You always have to be worrying about them. But you can manage if you have to," says one woman who has five children. The logistics of child care are demanding for all working women, but bay shrimping presents special difficulties. Rose solves her problem by taking her baby with her everywhere she goes. She has one set of baby equipment at her house, another at a bait camp near the dock where her husband ties up their boat; she carries additional baby supplies with her in their pickup truck.

For most bay shrimper families, housework is women's work. Women who assert that there are some things on a boat a woman just cannot do are also quite likely to proclaim that their husbands are of no use around the house; the women are protecting their gender territory. One woman sighs during her interview that she is a lucky woman if her husband so much as carries a coffee cup to the kitchen. In contrast, women who take more active roles on the boat are more likely to say that their sons or husbands also help at home.[12]

The majority of women tackle their housework problem by lowering their standards. Most of them live in trailers or houses with few frills. They have neither the time for nor the interest in cleaning and maintaining larger or fancier homes that some might easily afford, given their incomes. Several of the older women have typical grandmother houses, complete with abundant bric-a-brac, often with a fishing motif. They handle some of their housekeeping hassles by keeping the gear in the garage or a mudroom.

Although several women have gardens, few have time to tend them, and neither husband nor wife has time for much yardwork. As a result, yards are usually sparse, with grass and shrubs ill-attended; time is too precious to spend on watering and mowing. About a third of the women live directly on the water, with the boat tied up by the back door; for these families, repair and maintenance of the boat leaves little time to care for the exterior of their homes or the appearance of their yards. Women help with a variety of different kinds of boat repairs, depending on their skills. The men often discuss boat problems with them although the women's advice, even when sought out, is not always taken.

It is even more difficult for wives who live in trailers or pickups during certain parts of the fishing season; they find it next to impossible to keep an orderly house. Often the small living spaces are shared with other adults who shrimp. The trailers are so cramped and hot that the families spend much of their time elsewhere, returning during the cooler parts of the day and night to sleep and eat.

Female shrimpers, like their male counterparts, take a no-nonsense approach to their appearance while on the job. Yet many of them take considerable care to maintain their cultural definition of femininity in a male work setting. At work the women dress in functional and utilitarian clothes, favoring jeans and shorts. Like their husbands and boyfriends, the women get dirty and smelly working on the boats; cosmetics and fancy clothes are not found on a bay boat.

At home, however, the majority of women dress as expected of their generations. Older women are more likely to wear dresses, light makeup, and have styled hair. Younger women, in contrast, often wear shorts or jeans with a T-shirt, makeup, and personal accessories such as jewelry that clearly identify them as female. The women who captain their own boats wear clothing and carry themselves in a less traditionally feminine way, demonstrating both in physical appearance and demeanor that they can handle themselves in a man's world.

❖ ❖ ❖

DISCRIMINATION IN THE WORKPLACE

The women who were interviewed are quite aware of being females in a male world. Some, particularly those who have been shrimping for twenty years or longer, express their awareness as an acceptance of the likelihood of being teased by male shrimpers. They also recognize that they must stay away from traditionally male settings such as bars and fish houses. Aside from these limitations, older female shrimpers are confident of their position in a male work setting. "They had respect for me, all of them," says Mildred. "They'd calm down on the CB if they knew I was there."

But female shrimpers are expected to know their limitations, to know what is acceptable behavior and what is not. "If you act like one of those port city floozies, they'll treat you different," says one respondent. As Meg puts it, "A lot of women think shrimping is a man's place. And it is, unless you're with your husband." Meg's sentiments are echoed by the more traditional women who shrimp. As long as a woman is connected with a male fisherman and respects her gender-based limitations, she is treated "like one of the gang. The guys are real polite, respectful."

To be successful in a male world, one woman says, "you need to know the Texas way: kids first, then husband, then yourself. Honor and hard work." These values mesh well with conceptions of gender roles in traditional patriarchal societies where men fulfill the good-provider role. If a woman has young children who require her care, she provides for them under conditions that are at times stressful and demanding. Thus she may drop out of the work force while her children are young or just work summers with her husband, the children with their parents on the boat.

Husbands come next. Most women at first shrimp to earn extra family income or to spend more time with their husbands. Some women, like their male counterparts, take to fishing, grow to enjoy it, but others detest it and think of ways to disinvolve themselves. The work takes its toll; the women sustain the same kinds of minor injuries as their husbands, weather the same storms, endure the same repetitious and boring work. Nevertheless, women feel proud of their accomplishments and the ways in which they help their husbands and families; they feel they are living honorable lives.

In "the Texas way," personal needs come third and for most

women a distant last. This view does not mean that the women are unhappy or dissatisfied; it does mean that their lives are defined by the needs of their families, that attention to their own desires is a low priority. The second shift leaves little time and energy for personal wants; leisure time is scant, hobbies rare. There is little time either to rest or reflect.[13]

The women who were interviewed would not identify themselves as shrimpers per se but as shrimper's wives.[14] They were pleased to enlighten the researcher on how they juggled two tasks, family and fishing, but were puzzled that anyone would be interested in women who shrimp. The women, indeed, do not consider themselves as fisherman but as women helping their husbands, doing what women are expected to do. A woman who does not live by this culture code can be criticized heavily; sanctions serve to show her that she has crossed acceptable gender boundaries.

Darlene, who captains an all-male crew, believes that men are out to give her a hard time: "They like to see you get in a bind just to see what you can do." She was reluctant to call for help to other shrimpers on the CB because she "got so much shit." Instead, she calls her father to help her when she gets in a serious jam. Rose states that she "could outwork a lot of men" and adds, "They respect me, because they know I could outwork them." But she also feels that her hard work gets her in trouble with some male fishermen. "They're very chauvinistic. For that reason I'm not very popular with the guys." Rose also says that captains in her port berate their crews for being lazy or slow, comparing her hard work as a deckhand to their inadequacies as fishermen. She concludes that although she likes having a reputation as a hard worker, she thinks it is unfair that bay captains fling this fact in the face of their male crews and thus create animosity toward her. She feels a backlash directed against her and other female shrimpers. As a result, she says, "I try to keep a pretty low profile."[15]

The status of a female shrimper's husband also determines how she is treated on a boat and on the docks. One respondent, who is married to a captain who is not as successful as most, claims that male shrimpers look down on female shrimpers. "They call them lowlife," she says. "Really all you're trying to do is make a damn living. But they think you're trash. You can't put on makeup, dress up. Every time you get near the boat, you end up getting dirty. So you have trouble renting apartments."[16]

Some of the respondents in the survey believed that the prevailing values are changing. One woman in particular sees herself as a role model, breaking ground for women in the future who may choose to follow in her footsteps. In her estimation, the impact of the new gender identities that are being negotiated in other work settings is beginning to ripple through the fishing community, giving women who shrimp more recognition for their contributions.[17]

❖ ❖ ❖

POLITICS

In many fishing communities land-based women become political advocates for their seafaring husbands; only a few of the Texas women follow this pattern, however. Rose, for example, recognizes that political clout comes from organization and that men who shrimp do not have the time to devote to politics. Therefore she attends meetings and other activities that support their political interests. She also laments that most men are too bound by sexist gender expectations to accept women's contributions. "Men don't have the time for politics, but they won't give women the chance. Even if women did anything, men wouldn't give them credit for any of it. They would feel threatened."

Other women who shrimp claim their lack of political participation stems from fatalism about the political process rather than from gender limitations. "We've stayed on the numb side of that one," says one women, explaining why both she and her husband do not get involved. "We couldn't get together enough money to buy votes," asserts another. Some cite time constraints and the pressures of economics on bay shrimpers as reasons to keep out of politics.

"The same night of the meeting, the shrimp ran just well enough that nobody made it on time. We pay our dues, but don't have time to attend the meetings." Many others echoed Meg's fatalism: "It's a political game, all the rules and regulations are thought up for someone else."

Another female shrimper decries a political shortsightedness. Instead of working together with each other, she observes, bay shrimpers work against Gulf shrimpers and different ethnic groups work against each other, reducing the clout of bay shrimpers. Kate agrees. "The fishermen's their own worst enemy. They sit around and complain about changes that should be made, but then they claim they don't have the time or money to do anything. All they know is what they do. They don't think ahead."

❖ ❖ ❖

GENDER, FISHING, AND FARMING

The data from this subsample suggest that although women have been shrimping the Texas bays for at least half a century, perhaps longer, their names do not appear in public records, nor has their presence been documented in the research literature. The lack of recognition can be explained in part because the women have defined themselves less as commercial fishermen than as unpaid family helpers.

The career patterns of the women contribute to the self-definition of their work as nonfishermen. They often move in and out of shrimping, punctuating their working careers with time off to raise children or to run shoreside businesses. Based upon the cultural definition of employment, these periods in the unpaid labor force are not defined as part of their work history.[18] Women who shrimp are therefore often viewed (and view themselves) as dabblers who help their husbands for the fun of it, or in emergency situations, or from time to time to lower overhead.[19]

Women on ranches and farms in the United States historically have followed the same pattern. Because of the nature of ranch and farm work, there are frequent seasonal labor shortages. Wives and daughters have been expected to pitch in, just as bay shrimpers' wives, girlfriends, and daughters do on the Texas coast. Until recently, however, regardless of how regularly and actively they participated in day-to-day operations, rural women called themselves "helpers" or "farmers' wives" unless they had no husbands.[20] In a culture based on the ideology of "separate spheres," this attitude protected men's place as "good providers." Since women in rural communities defined themselves as helpers rather than as workers, society did not regard them as primary producers; consequently there has been little research on their contributions outside the home. Recent researchers have begun to discover women's roles as "girl homesteaders," as ranchers, and as farmers.[21] As women's definitions of their own identity change in the fisheries, research may reveal that women made a larger contribution than originally believed.

Women in American society meanwhile are moving into jobs that have been held by men; as this occurs, conceptions of gender are being redefined. Given this exploratory study, I would suggest that female bay shrimpers who most completely involve themselves in their work already view themselves as professional fishermen. For example, as Rose sees it, she and women like her provide role models for the other women who shrimp, especially for the majority of women who are still protecting their traditional gender identities derived from their ties to men. More and more women in Texas coastal communities may begin to redefine the work they perform on the bay and to think of themselves not only as wives and mothers but increasingly as bay shrimpers.[22]

Farming households are witnessing similar kinds of changes. Throughout the country, two-earner households have become the norm; few families can get by on one income, and rural families are no exception. Because men's incomes are generally larger than women's, it is often the husband who takes the off-farm job. When that happens, women are more likely to call themselves "farmers" rather than "farmers' wives" or "farmers' helpers."

In studying the place of women in corporations, Rosabeth Moss Kanter found that gender-based stereotypes were strongest when there were only token representatives of the opposite sex.[23] As more women joined the corporation, women, like their male counterparts, were judged on performance rather than on the basis of gendered expectations. If more women define themselves as fishermen or as farmers, it seems probable that they will be judged on the basis of what they do, not simply on their relationship to men.

8

Vietnamese Bay Shrimpers

❖ ❖ ❖ ❖ ❖ ❖ ❖ ❖ ❖ ❖ ❖ ❖ ❖ ❖

After the fall of Saigon, about 130,000 Vietnamese escaped to the United States.[1] Additional waves of war refugees continued to arrive, and by 1979 approximately 220,000 Indochinese had been resettled in this country. A small portion of the new immigrants settled along the Texas coast and, according to one Texas bay shrimper, were "taking over shrimping."

The Vietnamese and other Indochinese who migrated to Texas came from various backgrounds although they shared certain commonalities. Some who fled to the United States had originally escaped from North Vietnam after the Communist takeover in 1954.[2]

The majority of fishermen who eventually were resettled in Palacios and Seadrift, however, were poorly educated peasants who had fought as soldiers in the South Vietnamese army; among these men were also a few low-level government administrators. Vietnamese fishermen and their families in Palacios were from the village of Nam Dinh in the province of Ha Trai; those who came to live in Seadrift were from a number of other small villages, the majority on or near the Vietnamese coast. Notable among the small group of Laotian fishermen who settled in Seadrift were the Hmong, mountain villagers who had fought against the North Vietnamese under the aegis of the Central Intelligence Agency. The majority of Vietnamese fishermen in Bayou La Batre, Alabama, several of whom had recently moved from the Texas coast or were returning to it, were from small coastal villages bordering the Gulf of Thailand.

Before their relocation, refugees were first funneled into four camps in the United States; their resettlement was accomplished through voluntary groups and organizations. No consistent federal policy determined where or how the immigrants were relocated after they left the camps. American churches took an active role, and Catholic parishes were especially instrumental since many of the refugees were Catholics.

The majority of Vietnamese immigrants were resettled in California although they freely moved from one part of the country to another to seek better employment and to be near other Vietnamese refugee communities. They were also resettled in the states that bordered the Gulf of Mexico.[3] About 20,000 were relocated to Texas, including those who found new homes along the Texas coast.

Voluntary organizations, churches, and the federal government believed that the Gulf states were desirable destinations because they closely approximated the climate of Vietnam. It was assumed that the men would soon find work in commercial fishing, employment in which some had experience in their homeland. Given these beliefs, however misguided, the Texas coast was a logical place for the refugees' resettlement.

Texas bay shrimpers did not set out the welcome mat for the newcomers; instead, they put aside their long-standing animosities toward Gulf shrimpers to combat the "outsiders." Yet their response should not be considered simply an expression of racism, conveniently fitting the stereotype of the Texas redneck. The reaction of bay shrimpers to the Vietnamese was complex, brought about by a combination of their own sensibilities and the actions of the Vietnamese themselves. The story of the resettlement of the refugees reveals much about the politics and the economics of shrimp fishing and about the lives of commercial fishermen within their own communities.

❖ ❖ ❖

PORT ISABEL

Vietnamese refugees eventually were resettled in several coastal locations in Texas, including Seadrift, Palacios, Seabrook, Sabine, Fulton, Rockport, and Port Isabel. The problems of the immigrants who were resettled in Port Isabel exemplified from the start some of the conflicts that ultimately emerged along the Texas and Gulf coasts.[4]

About 150 Vietnamese men, women, and children moved into Port Isabel, a poverty-ridden coastal community situated in one of the poorest regions in the United States.[5] The Diocese of Brownsville facilitated the resettlement process in Port Isabel in 1975, which at the time had a population of about 3,000, 85 percent of whom were Mexican American, the balance Anglos. The town's major industry was Gulf shrimping although tourism on South Padre Island, just across the Laguna Madre, was increasingly a mainstay of the local economy; there was no significant bay shrimp industry in the Laguna Madre.[6]

Vietnamese men and women were hired by the owner of a fish house and fleet. The men worked as crew on Gulf shrimp trawlers, the women sorted and packed shrimp in the fish house. Soon the Vietnamese workers complained about the low wages they were paid, the working conditions, and the bad treatment they felt they received from the residents of Port Isabel. Not only were they being exploited economically, they asserted, but they also were being discriminated against because of their race.

Complaints turned into more serious hostilities between the new immigrants and the established residents of Port Isabel. The Mexican American majority and the Anglo minority in the community did not understand the language, the culture, or the traditions of their new neighbors. Moreover, the Vietnamese seemed to the Texans to be unwilling to learn English and if they did understand it to pretend otherwise. A number of charges by both groups were hurled back and forth. The local residents accused the Vietnamese of unsanitary health practices, which included washing their clothes in the boat canals where tourist fisherman cast their lines. The Mexican Americans and Anglos felt that the Vietnamese were making no attempt to fit in, actively resisting the process of Americanization that was central, from their viewpoint, to acceptance and success. In turn, the Vietnamese accused the fish house operator of gross exploitation and of general neglect of their welfare; they believed that they were being taken advantage of because they were refugees.

The Catholic Diocese of Brownsville attempted to act as an arbitrator but without much success. The tensions between the two sides escalated, both Americans and Vietnamese predicting possible violence. In early 1977, after less than a year, the 150 Vietnamese immigrants packed their bags and left Port Isabel. Members of the community rejoiced that they had gone.

❖ ❖ ❖

SEADRIFT

About 100 Vietnamese immigrants were resettled in Seadrift in 1975.[7] From the beginning the newcomers were not welcomed by the majority of the town even though many of them had fought along with American troops against the North Vietnamese Communists.

Hard economic times were a determining factor in the treatment of the Vietnamese, but the locals also saw them as a direct threat to the major industry. Seadrift, a coastal community of about 1,200, had relied since its earliest beginnings at the turn of the century upon the bay fishery to provide its major employment.[8] About 100 bay shrimp boats were based in Seadrift, along with nine small fish houses located near the dilapidated docks. Oysters had been a mainstay of the local economy for many years, but by 1975 the oyster beds in Matagorda Bay had gone bad. Shrimp prices were down. Fishermen, and the community businesses that relied upon them, were not doing well. In fact, many local shrimp fishermen believed that the bay already had too many boats and that the catch per vessel was on the decline because of the excess. Therefore, they reasoned, the last thing Seadrift needed was refugee fishermen.

To make matters worse, several nearby chemical plants and refineries, including the Alcoa plant in Port Lavaca, cut back their labor forces because of a general downturn in the national economy. Some of the bay fishermen, because of hard times on the water, had taken full- or part-time jobs at the plants to tide them over until the shrimping improved. With the layoffs and poor fishing, Seadrift faced hard economic times, and no immediate solutions were in sight.

In many ways Seadrift was no different from other small and isolated coastal communities up and down the Texas coast. A politically conservative community like Port Isabel, Seadrift put a high value on the work ethic and the importance of family and religion. Residents of Seadrift, in contrast to Port Isabel, were predominantly white; minorities composed less than 1 percent of the population. The town had sent more than its fair share of soldiers to Vietnam, and negative feelings about the war in Southeast Asia still lingered. The defeat of American soldiers at the hands of the North Vietnamese, the body counts, and the wounded were still fresh in the minds of many people in the community who had sent their sons, brothers, fathers, and uncles to the controversial war. Many residents still retained a bitterness about the conflict that only time could heal. As one resident put it, "Everyone here just wanted to forget it."

The refugees in Seadrift were treated badly from the start. They were seen as a major threat to the economic welfare of local residents, as competitors on the bay. Their presence also stirred up war memories and anger. More than a few residents of Seadrift were confused about who the refugees were, believing them to have been aligned with the North Vietnamese and thus directly responsible for the war deaths and the American defeat. Racism also played a part; for example, the new refugees were called the same names that American soldiers had called them in Vietnam, including "gook" and "slope."

For their part, and adding equally to the trouble, the new refugees were often insensitive to the community's traditional informal and formal rules that governed bay fishing. First and foremost, they fished every hour of the day. The new fishermen took few breaks from their work and always seemed to be out on the water, even in bad weather when the Texans stayed tied up at the dock. They appeared to work round the clock, twenty-four hours a day, every day. The Texas shrimpers referred to the practice as "clocking." They did not take it lightly. From their perspective, the refugees were systematically overfishing the bay, a practice that eventually would translate into lost income and lost jobs.

The Texans' suspicions found support in their observation that the Vietnamese did not throw any of their catch back into the bay. Seadrift fishermen swept the trash fish over the side of the boat after it had been separated from the shrimp; the Vietnamese kept everything. What the new refugees could not sell for a few pennies, they gave to family members. They consumed species of fish that Texas shrimpers would never dream of eating. Locals viewed this behavior not as a cultural curiosity but as another sign that the Vietnamese were "barbaric" in their customs and traditions. Seadrift fishermen claimed that the Vietnamese were permanently damaging the bay, that it was just a matter of time before the shrimp would be gone.

A number of stories made the rounds in tiny Seadrift, tales that superficially confirmed the strange customs of the new refugees. One rumor charged that the Vietnamese captured sea gulls and other seabirds and ate them. Several Texas shrimpers told a story about how the Vietnamese nailed the feet of wild birds to the decks of their boats, cooking the birds when needed. The stories were told as confirmation that the Vietnamese were "like animals" and "not like you and me." Although nailing birds to the deck was uncommon among the Vietnamese fishermen, several different respondents—two Vietnamese fishermen in Bayou La Batre, one in Seadrift, and two in Palacios—confirmed that some Vietnamese shrimpers did do so when they first came to the Texas coast.

In addition to clocking, the Vietnamese often ignored limits on

A bay captain ties off a line on the work deck while dragging on Galveston Bay. The live shrimp well is in the foreground.

catch, gear restrictions, and other rules that governed the bays. Texas bay shrimpers, of course, systematically broke the same rules. Vietnamese also were accused of not following other informal expectations that the Texas shrimpers considered important. Particularly irksome to the local shrimpers was the fact that if the Vietnamese spotted a Texas fisherman who hit upon a good run of shrimp, they might rush to the same spot; from the local shrimpers' viewpoint, they were stealing the Texan's shrimp—or the shrimp his friends might have netted after he radioed his location to them. Seadrift shrimpers observed that Vietnamese fishermen pulled their nets in the "wrong" direction and piloted their boats dangerously close to Texas boats. In short, the residents of Seadrift believed that the Vietnamese were overly aggressive, sometimes dangerous, and were destroying their livelihood.

A further grievance was that Vietnamese fishermen clogged up the marine radio band with ceaseless and senseless chatter, according to the Texas shrimpers. The locals were used to long conversations with each other, including trading information about weather, the best fishing spots, and a variety of other topics. They also listened in on other shrimpers to see if they could find out where the best fishing was taking place. But eavesdropping on the Vietnamese was a waste of time; the refugees never spoke English and few locals spoke or understood even a little Vietnamese.

The Vietnamese were good at playing dumb when the TPWD agents pulled them over for breaking a regulation. The local shrimpers claimed that if caught red-handed, the refugees would pretend they did not speak English and could not read the TPWD handbook on regulations. In contrast, in the cat-and-mouse game between Texas bay shrimpers and the agents, the assumption was that once a shrimper was caught, he admitted his guilt and paid his fine "like a man should."

Seadrift shrimpers grew even more angry when it became evident that the Vietnamese were making good money at shrimping while many longtime local fishermen were just barely getting by. The Vietnamese soon bought or built new boats, purchased trucks and vans, and moved into their own homes. The lifelong residents of Seadrift were certain that at least some of this affluence was from welfare payments, and they also believed that the refugees received special grants from the federal government. Some Vietnamese, too, were thought to be cashing in a few of the gold bars they had brought with them from Vietnam. Many shrimpers in Seadrift said the Vietnamese had large amounts of gold with them that they hoarded behind their public face of austerity and need. The hidden gold explained how the new refugees were able to pay off the boats and purchase new vans and buy electrical appliances. A final story, one for which I could find no substantiation, was prevalent up and down the entire Gulf of Mexico coast, wherever the Vietnamese had settled. It had several variations but always explained how the pet population in a particular neighborhood would decline drastically after a Vietnamese family moved in: the refugees were accused of stealing family pets at night and eating them.

Hostility between the longtime residents of Seadrift and the refugees steadily increased over the course of the first year. Then quite suddenly the Vietnamese who owned bay boats purchased at relatively high prices from local fishermen moved north to Seabrook and Palacios, or south to the Rockport and Fulton areas.

The Vietnamese who remained in Seadrift went to work for the Bo Brooks Company, the supplier for a seafood restaurant in Baltimore, Maryland, that offered its customers crab from Texas bays. The small crab plant, a drab metal building on the edge of town, soon employed almost all the Vietnamese women who remained in Seadrift. While the women picked crab for Bo Brooks, their husbands, brothers, and fathers entered the local crabbing industry, harvesting crabs from baited pots placed in the shallow bay.

By all accounts the Vietnamese women were excellent crab pickers. They earned a piece-rate based on the pounds of crab they separated from the shell and packed in small plastic containers to be shipped to Baltimore. The fastest picker earned more than sixty dollars in an eight-to-ten-hour shift. The women stood at long white counters for hours at a time picking crab with skillful hand movements that required excellent hand-eye coordination as well as concentration. The women often skipped their coffee breaks to continue working, to earn extra dollars. The plant manager told me that the Vietnamese women were the best crab pickers he had ever seen.[9]

The Vietnamese men quickly learned how to fish for crab. The fishery required a smaller investment of capital than bay shrimping; it was also a far easier job to learn. Texas crabbers worked their baited pots from small skiffs powered by gasoline outboards. At first the Vietnamese men in Seadrift did not know the best places to crab and so mimicked the Texas crabbers, much as they had the bay shrimpers, placing their pots near or in the same spots that were used by the handful of resident Seadrift crabbers who worked the local bay. The latter saw the refugees as intrusive, crude, menacing, a threat to their livelihood; at worst, they were perceived as thieves, taking pots and stealing crabs. Once again the Texas fishermen charged the Vietnamese with breaking both formal and informal practices in fishing. The Vietnamese crabbers, according to the resident fishermen, overfished, fished out of season, and broke every possible rule. Furthermore, when caught by the TPWD agents, the refugees pretended not to understand English and pleaded ignorance of the fishing laws. It was the same story all over again.

Although the Seadrift crabbers usually exaggerated the facts in their stories and observations of the new refugees, there was some truth to their claims. The Vietnamese crabbers and shrimpers were accustomed to their own methods in Vietnam, where they fished with minimal gear and no regulations. Inshore fishermen there used small skiffs powered by old gasoline engines that often broke down; when their engines quit, they took out their oars and rowed back to shore. They fished a variety of species, none of which was governed by catch limits or other regulations. When their catch was landed at the dock, it was transferred by hand to straw baskets and quickly carried to market; ice was rarely available.

Before the war disrupted commercial fishing in Vietnam, old wooden trawlers plied the deeper waters of the Gulf of Thailand, crewed by eighteen or more men. These offshore vessels had to return to land frequently to take on food for the crew and fuel. As was the case with the inshore fishery, there were neither laws to govern the fishing nor agencies to enforce them. When I asked one refugee shrimper who had fished for many years in Vietnam what happened when his vessel broke down at sea, he shrugged his shoulders and looked puzzled. I had assumed that the captain of the boat radioed for help to the Vietnamese navy or Coast Guard. "You wait," he told me, until a ship comes by. "Maybe they help you. Maybe they don't."

Even though the Vietnamese had withdrawn from the bay shrimp fishery, tensions in Seadrift mounted as stories about the Vietnamese crab fishermen made the rounds. Bay shrimpers were further angered when they heard about problems with Vietnamese shrimpers on Galveston Bay. The Seadrift fishermen heard firsthand and secondhand accounts from other bay shrimpers, which they then passed along to their friends, that the Vietnamese were clocking on Galveston Bay and also engaging in other fishing practices deemed deviant by local standards and customs.

Then the local shrimpers began receiving sympathy and support from Texas Gulf shrimpers, who feared that the Vietnamese, if successful in the bays, would soon fish the Gulf of Mexico. Putting aside their long-held distrust and hatred, the two groups of fishermen banded together against a common threat, the newcomer refugees.

In Seadrift and elsewhere along the Texas coats, there was serious talk of teaching the Vietnamese a lesson, of burning their boats to show that they were unwelcome, that they should leave Texas and "go back where they came from." It was presumed that such a public display would intimidate the refugees. If threatened, perhaps the refugee crabbers and their families would leave Seadrift, too, finally ridding the community of the immigrants.

Hindsight suggests that the shrimpers in Seadrift and elsewhere in Texas were naive to assume that the Vietnamese would respond to intimidation and threats, public or otherwise. But the Texas shrimpers knew little about the refugees and failed to realize that most of the Vietnamese men had been soldiers for many years, that they and their families had endured years of war and hardship in their native country. The Texas fishermen simply presumed that threats and boat burnings would make the newcomers leave town for good.

❖ ❖ ❖

THE DEATH OF BILLY JOE APLIN

One Seadrift crabber, Billy Joe Aplin, had reason to be especially angered by the Vietnamese fishermen.[10] He had been crabbing in the bay with his family when one of the Vietnamese began placing pots next to his. According to Aplin, he yelled at the intruder to go away, but he soon returned with several other boatloads of Vietnamese fishermen who then verbally abused Aplin and his family and made threatening gestures. Two suspects were arrested, but Aplin did not press charges and they were released.

Aplin was still angry, but he also feared for the safety of his family; he believed that to press charges would further antagonize the Vietnamese. He decided to get out of crab fishing; he would sell his crab boat and start shrimping in the bay. On August 2, 1979, he was repairing his boat with a welding torch, getting it ready to sell, when he seriously burned his eye. His wife took him to the hospital, and when he returned home he slept the rest of the day.[11]

On his way back to the dock the next day, Aplin passed by two of the men who he believed had harassed him while crabbing. Aplin got out of his truck, approached one of the men, Nguyen Van Sau, and hit him. Some witnesses said Aplin beat Sau ruthlessly, others that he struck the refugee only once. Sau and his brother, Nguyen Van Chinh, jumped in their car, went home, and got a gun. When they returned to the docks, Aplin allegedly threatened both of them with a knife. Nguyen Van Sau shot Aplin five times; he died from the wounds. The two Vietnamese were arrested and charged with murder.

That night the home of a Vietnamese family in Seadrift was firebombed and burned to the ground; several Vietnamese boats were also burned. Although no one else was hurt, the refugees understood the seriousness of the situation; the majority of the remaining Vietnamese in Seadrift packed their bags and immediately left town.

The media focused on the killing of Billy Joe Aplin, and the story received major coverage in the national newspapers and on the network news. ABC's "20/20" devoted a segment of its weekly program to a report on the circumstances surrounding the death. Papers throughout Texas detailed the killing, boat burnings, and subsequent trial of the two defendants. Director Louis Malle produced a Hollywood film based on the conflict in Seadrift, *Alamo Bay*.

The Ku Klux Klan, hoping to attract support and publicity for its own political agenda, entered the picture even before Aplin was killed. The avowedly racist organization offered to sponsor a protest on the streets of Seadrift and to burn a representation of a Vietnamese boat. The national media portrayed Seadrift as the home of 1,200 racists, as did the Hollywood actors in Malle's film, but few residents of Seadrift supported the Klan. The majority viewed its members as opportunistic outsiders and would have nothing to do with them.

The U.S. Department of Justice attempted to mediate the dispute between the fishermen after Aplin's death. A study was commissioned by the Minority Business Development Agency of the USDC that focused on nonfishing jobs that the Vietnamese might find along the Texas coast.[12] Various state officials were called together to devise a plan to resolve the problems in Seadrift and in the other coastal communities where the Vietnamese had resettled. The Texas governor initiated a task force. Seadrift residents, both Texans and Vietnamese, temporarily formed a group, the Seadrift Community Council, as "a start toward settling things that have happened here or could happen in the future."[13]

Tensions remained high in Seadrift throughout the trial of the two Vietnamese accused of murdering Aplin, and the national media kept the incident fresh in the minds of many Americans. Aplin's mother, interviewed on ABC's "20/20," said, "It is a sad occasion to think that you have raised eight children and your oldest is 44 . . . [and your] youngest son was 35 when he was killed in America. He didn't have to go to war. He was killed here by the Vietnamese. Not over there."[14]

Another established resident of Seadrift said after Aplin's killing, "What's to keep a man from going out there and shooting one or two [Vietnamese], throwing them in the water? You watch and you'll see the next year it's going to happen. If they don't stop it, the people here will stop it."[15]

A Matagorda County jury found the two Vietnamese innocent of murdering Billy Joe Aplin. The jury reportedly believed that the two men had reason to fear for their lives and were defending themselves. After the trial, with the exception of the filming and release of *Alamo Bay*, the media dropped the story. The mediator from the Justice Department left town. The governor's task force disbanded within a year, viewed by some of his critics as a political sham to make him appear as if he were actively engaged in community problemsolving. The study commissioned by the federal government recommended a variety of solutions to the problems in Seadrift, including the development of new marine-related industries, the training of coastal community residents to work in them, and the education of Texas and Vietnamese residents to help them be more sensitive to the needs of the others' cultures. But there were no funds to implement the recommendations. Several months after the verdicts were handed down, the Vietnamese crabbers and their families returned to Seadrift. The men resumed their crab fishing, the women their work in the crab plant. Vietnamese shrimpers and their families did not return, however, still concerned for their safety.

By 1984 there were some definite signs that the Vietnamese had received a limited acceptance in Seadrift. Violence between the two groups had subsided although the relationship between Texas bay fishermen and Vietnamese crabbers remained rocky. One indication of progress that residents often mentioned was that the salutatorian that year at the high school was Vietnamese.

Perhaps an even more appropriate sign of growing community tolerance was the response of the locals to a move by the Vietnamese women at the crab plant. In summer 1984 the women went on strike for higher wages; they also demanded higher prices for the crabs their husbands, fathers, and brothers sold to the plant. The plant manager was both surprised and puzzled by the turn of events, but many townspeople sympathized with the Vietnamese strikers, who eventually won their demands.

The mayor of Seadrift in 1984 was optimistic about the future of the small town. He expressed a common sentiment that although the community had confronted severe problems, including adjusting to the refugees, things were looking up. The Vietnamese were no longer blamed for the hard economic times that residents had experienced.

Deep-seated mistrust and fear between the two groups did not disappear, however. Vietnamese fishermen were not allowed by local fish house operators to dock their boats where everyone else did; by 1989 the Vietnamese were still prohibited from parking their boats beside the Anglos'. Although overt hostilities subsided, the refugees were in many respects second-class citizens living in a poor, isolated fishing community. The charred remains of the Vietnamese house that had been burned on the night Billy Joe Aplin was killed were covered by high grass and shrubs. Although the Vietnamese were allowed to live in relative peace in Seadrift, few residents wanted them as neighbors.

❖ ❖ ❖

THE FREEZE ON LICENSES AND LIMITED ENTRY

Texas and Vietnamese bay shrimpers in other coastal communities along the Gulf of Mexico reacted to the death of Billy Joe Aplin and the events surrounding it in ways that were often played out far from the political arena. Response from fishery managers and politicians, however, centered on formal policies and regulations that could resolve the conflict between the two hostile groups as the Texas legislature took up the issue of limited entry to the bay fishery. The shrimpers themselves, in general strongly opposed to any kind of limited entry plan, attempted through informal agreements to resolve their differences.

The Anglo and Vietnamese bay shrimpers on Galveston and San Antonio bays devised a written agreement to limit the number of new boats coming into the fishery. For example, on Galveston Bay the latter agreed in part "to discourage any Vietnamese against moving into Seabrook or buying any more boats. The Vietnamese agree to sell their shrimp for the same price as the native shrimpers or within 10 to 15 cents of that price. The Vietnamese would also not drag one net with two boats." In Palacios, south of Seadrift, Vietnamese and American shrimpers, anxious to ward off any attempts by the Texas legislature to impose limited entry, signed a "Statement of Consensus" in May 1980.[16] Although these and other such agreements along the Texas coast did much to ease tensions temporarily, they were neither binding nor strictly legal and could easily have been challenged in a court of law. However well intentioned, the temporary truces soon fell apart.

The bay shrimpers' response to suggestions by the legislature that limited entry be considered was particularly volatile in places such as Seabrook, where there were seventy bay boats owned by Texas shrimpers, fifty-five by Vietnamese fisherman. One Seabrook shrimper pointed out, "We break all the rules and the Vietnamese don't do it anymore than the Americans. But the law's got to do something to stop this."[17] Angry about the growing number of boats on both sides, the fisherman still distrusted the legislature to find a workable solution.

The experts, representing the interests of the federal, state, and local branches of government, studied the problems between native and refugee shrimpers in Seadrift. Concluding that the conflict was the result of a scarcity of economic resources, they completely discounted, or deemphasized, the importance of race. The state legislature, which then regulated the bay shrimp industry, thus sought economic solutions to the problems in the fishery that could be turned into law. Data were assembled that demonstrated that declining catches were directly attributable to a rise in the number of boats in the bay fishery. The legislature, following the advice of the TPWD and other state programs and agencies involved, put a freeze on additional bay licenses in 1981. The intent of the legislation was both to allow time for the fishermen's tempers to cool down and for an acceptable limited entry plan to be developed. The ban was viewed as temporary.

Other problems in the bay shrimp industry were actively ignored, including those stemming from cultural and racial differences. The misunderstandings among fishermen that had surfaced at Seadrift certainly were present in the other communities where the refugees had been resettled. Moreover, the issue of falling prices for shrimp—the result of increased foreign production—was avoided.

In 1983 the state legislature, responding in part to the lobbying efforts of bay shrimpers, decided to remove the ban on bay fishing licenses, and serious consideration of limited entry was abandoned. The legislature took the position, along with the TPWD, that the market would quickly eliminate any marginal bay shrimpers, both Vietnamese and Texan, which in turn would increase the catch per boat. Limited entry was redefined as a bogus attempt to interfere with the free market.

The ban on licenses and the serious talk about limited entry had resulted in one positive effect on the hostilities between the two groups of bay shrimpers by allowing them to hate a common enemy, this time the legislature. The ban on new boats was beneficial in that it created a cooling-off period, allowing both sides breathing room to go about their daily work of fishing the bays.

Bay shrimpers no longer made national or state headlines, but hatred between Texans and Vietnamese was not far beneath the surface. Racism and cultural ignorance did not disappear after the killing of Billy Joe Aplin, nor did the Vietnamese stop the clocking and the other customs that infuriated Texas shrimpers. There were claims by both Texas and Vietnamese fisherman that shots were fired on the water. The accusations were hard to substantiate since law enforcers were rarely in the area at the time of the alleged gunplay. It was common knowledge that both sides were heavily armed.

❖ ❖ ❖

PALACIOS

Events unfolded in Palacios in a slightly different way from those in Port Isabel and Seadrift.[18] The new refugees who were settled in Palacios, a town of several thousand that sits directly on Matagorda Bay between Port Lavaca and Freeport, were different from the immigrants who lived in either Seadrift or Port Isabel because they came from the same village in Vietnam. As in their country of origin, they remained under the direct leadership of their council of elders. Moreover, the same village priest who had led them in their faith in Vietnam now taught them in Texas. He had been educated by the French and spoke English very well; from the beginning the small refugee community in Palacios had an articulate public spokesman who was a quick study in American community politics and power.

The new refugees in Palacios acted as a cohesive group, the elders making decisions and formulating policies that were passed along to the rank and file; the priest served as public spokesman and dealt with the white community. The Vietnamese presented a united front in Palacios, a refugee group that shared its limited resources, learned from its mistakes, and, when necessary, aggressively attempted to confront the racism it encountered. Conflict between the refugees and the local residents generally was played out in public debate within the system or Palacios' custom and law. Violence was largely avoided because of the skills of the Vietnamese and the white community leaders, who were more sophisticated in handling conflict than leaders in either Port Isabel or Seadrift.

A bay boat from Palacios fishes on Matagorda Bay.

The first small group of refugees settled in the northern part of the community at some distance from the residential areas of Palacios. Unlike the newcomers in Seadrift, the Vietnamese in Palacios chose from the beginning to segregate themselves, physically to live apart from the townspeople. Their decision was partly a defensive posture because the refugees expected trouble, but it was also an attempt to avoid being seen as an overt threat; it made the statement that they wanted to keep to themselves. The new refugees signaled from the beginning that they would be good citizens of the community, but they were to be left alone.

The Vietnamese bought or rented used trailers and mobile homes, often sharing the small structures among two or more extended families that might number more than ten individuals. They pooled their money, as did the Vietnamese in Seadrift, and, as soon as they could afford to, made small additions to their trailers, most often extra space for bedrooms. The modest additions soon became a matter of contention between the refugees and the citizens of Palacios, however.[19] The issue itself became secondary, but it served as a focal point for airing the feelings of the shrimpers and other residents of Palacios. The conflict did not take place at the docks, on the water, or in the streets of Palacios but was largely restrained to a public forum, debated in the city council and on the editorial pages of the local paper, the *Palacios Beacon*, which had a circulation of 2,000 and was widely read throughout the small community.

Although shrimping had always been important to the economic livelihood of the community, it was not the only game in town; there was a nuclear energy plant nearby and a large Alcoa aluminum plant within commuting distance. Palacios had community leaders with cooler heads than those in Seadrift, some of whom could see that the Vietnamese could be an asset to the small community and who were determined to fight racism and discrimination within their own ranks. Nevertheless, some residents wanted to deny the allocation of building permits in order to limit additions and thus intimidate and harass Vietnamese fishermen and their families.[20] Several city council members, spurred on by residents, complained about a variety of problems that the Vietnamese were supposedly creating when they built additions to their tiny homes.

The city council and the *Palacios Beacon* became the public forums for a debate that went on for months. Rumors and myths that fueled racial hatreds in communities such as Seadrift were confronted directly in Palacios. The newspaper closely followed the city council debates, raising the level of dialogue on its editorial page from shouts and curses to position statements and rational argument; its letters and editorials aired fears, presented different sides of the issue, and through the pressure of public will and consent ultimately contributed to negotiation and compromise.

Rumors abounded in Palacios, as in other Texas coastal communities, about how the Vietnamese fishermen achieved their affluence. Most townspeople in Palacios, however, knew some of the truth about the refugees because of stories that appeared in the *Palacios Beacon*. For instance, the newspaper reported that the Vietnamese priest had arrived in town in 1976 after visiting many of the other coastal communities in Texas and had little money; he had slept in the back seat of his ten-year-old car to save on motel bills. The priest told me that he had put himself through a college English course at the University of Houston before he set out to find a home for his people. Lacking educational funds, he had applied for a Basic Opportunity Grant to pay for his tuition. The course had also allowed him to learn more about American culture and politics.

In June 1976 the priest brought six Vietnamese families to Palacios; they had almost no money, and there were no jobs waiting for them. One man was hired as a janitor by the South Texas Nuclear Project; six families lived on his income for several months because no one else could find work. Then several of the women were hired at the local crab plant where, as in Seadrift, they worked long shifts picking crab. For their labor they earned as much as $100 a day, which was apportioned among the six families along with the janitor's income.

The Vietnamese men were gradually able to buy carpentry tools and lumber, and they began to build their own boats, modeled after the boats they saw tied up at the Palacios docks. The men had been fishermen in Vietnam before they were drafted into the army as soldiers. Other Vietnamese who came later built their own boats or bought old boats from longtime residents; the refugees were always quick to point out that they paid a high price to the Americans for boats that often needed extensive repairs before they could be fished. After the first refugees had built their fishing boats, they learned how to shrimp. The priest told me, "Vietnamese can watch and learn. They are good at it. They look at something one time, then they learn how to do it." The Vietnamese watched the Texas bay shrimpers fish on Tres Palacios Bay, and through trial and error and much hard work, they gradually became good shrimpers.

Under the direction of the council of elders more and more Vietnamese families resettled in Palacios. Soon there were a total of 600 Vietnamese, about 100 families. There were also an additional twenty or so Vietnamese who came to Palacios who were not from the same village in Vietnam, not Catholic, and who did not live with the other refugees.

The Vietnamese made good money from their fishing, pooling their resources, buying better boats and equipment when they could afford it. The women who could worked in the crab plant, where they received a high wage compared to what they could earn in Vietnam. Their earnings were commonly shared and loaned among refugee families for the good of the Vietnamese community, the elders often supervising the financial dealings. At no time, the priest told me, did the families ever receive any government payments or live on welfare.

The lives of the new refugees, while promising, were not without tragedy. Two Vietnamese fishermen died from work-related accidents during their first three years in Palacios, one from exposure when he fell overboard, another who drowned when his boat overturned in the bay. Both were crabbers, although the majority of men in Palacios were shrimpers. One six-year-old boy drowned while playing on a shrimp boat at the dock.

When the refugees first moved to Palacios in 1976, they lived in

Bay Vue Mobile Home Park on East Bayshore and rented the small trailers; later they were able to buy used ones that were in better condition. The trailer park was convenient because there were docks just across the street fronting Tres Palacios Bay. The Vietnamese liked Bay Vue because of its isolation; they were left alone. The trailer park was separate from the town, its only approach a narrow asphalt road that crossed a small canal bridge, which was soon called "Chink Bridge" by some residents in Palacios. According to the priest, his people desired only to be left alone to earn a living and to practice their faith and way of life as they knew it.

The priest worked out a deal with one of the local Catholic churches. Every Sunday the Vietnamese used the church to celebrate mass in their language, but as soon as the service was over, the refugees hurried back to their homes. The priest dreamed of having his own church where his flock could worship as they had in Vietnam, and the refugees saved their hard-earned money for a new building.[21]

Complaints about the additions to the refugees' trailers finally peaked in 1982. The Vietnamese were puzzled, failing to understand why some citizens of Palacios were so concerned about their affairs. The letters to the editor streamed in to the *Palacios Beacon*; residents and several councilmen spoke out against the Vietnamese at the city council meetings. A committee was formed to study the possibility of creating new zoning laws in Palacios that would make it illegal for the Vietnamese to continue adding on to their trailers; the issue was referred to as the "permit problem." The insightful Vietnamese priest understood; the real problem, according to him, was that one of the city councilmen was "a rich man who also owned a trailer park. This man had trailers in his park just like ours with additions to them. He was just discriminating against us."[22]

When the Palacios City Council met in early May 1983 to make a final decision about the "permit problem," a new charge was aired: the Vietnamese were accused of building boats that could start serious fires. A citizen complained, "With the rosin and fifty-five-gallon drums, there could be a catastrophe in this town if a fire got started and we would not be able to control it."[23] The mayor responded that he was worried that the town's fire department would not be able to put out such a fire. The council then considered requiring various permits to limit boat building.

The mayor also said he had decided to require permits for the new additions to trailers in Bay Vue Trailer Park. Yet he realized, he said, that if the city required permits only for the new additions, then it would have no control over how the previous additions had been constructed. This situation, he continued, could affect the next renters when the Vietnamese moved somewhere else. One councilman, sympathetic to the Vietnamese, said, "It might be better just to ignore it and not take any action since the problem has existed for two years. [Let's] just hope they find other property."[24]

Amid heightening tensions, the Vietnamese elders decided that the best alternative was to buy land outside of town but still within the city limits and its services. The land the elders looked at was expensive, however, as was the cost of development, which included installing paved streets, sewers, and other infrastructure necessary to meet the city codes. An agreement was negotiated between the Palacios City Council and the Vietnamese elders, the priest serving as their representative. The city allowed the refugees to move their trailers from Bay Vue to the new development, Bethany Park. They were given up to five years to replace the trailers with homes, a special relaxation of building codes for the Vietnamese. Refugee families who could not immediately afford to build homes to code could continue to live in their trailers on their new lots until they could afford to build.

Meanwhile, a group of white bay shrimpers met in Palacios to lend their support to the United Shrimpers Association (USA) of Texas.[25] It advocated that bay shrimpers begin to control and limit the number of bay boats that fished in Texas, a form of limited entry. The shrimpers were certain that if the industry did not regulate the number of bay shrimp boats, then the legislature would do it for them. Representatives of the Vietnamese community were present but did not sign up to join the USA; limited entry was clearly not in their best interests. At the time they were building between twenty to thirty new boats, and they viewed the USA as a white-controlled organization whose purpose was to curtail their success in bay shrimping. On their part, the local shrimpers viewed the Vietnamese fishermen's failure to join the USA as another indication of their hostility to bay shrimpers and to the citizens of Palacios.

On June 15, 1984, eight years to the day after the six refugee families had moved to Palacios, groundbreaking began at Bethany Park, located just off Highway 35, several miles distant from other neighborhoods but within the city limits. The twenty-five acres were divided into seventy-nine lots. The mayor and other members of the city council made speeches and expressed hope for a better future for the Vietnamese and for the town of Palacios.[26]

The refugees immediately began moving their old trailer homes to the new development, and those who could afford to began construction of new houses. Four broad, straight streets were laid out, and sewers, lights, and other necessities soon followed. The Vietnamese built two- and three-bedroom houses for themselves, often with their own hands. Several of the new houses were elaborate; it was not long before some residents in Palacios made negative comments about the "mansions" in Bethany Park. Once again the rumor spread that the Vietnamese had hoarded their welfare payments or had cashed in gold bullion that they had brought with them from Vietnam. Problems between the two groups of shrimpers diminished after the construction of Bethany Park but did not completely disappear. Among other complaints, the white shrimpers continued to claim that the Vietnamese clogged the marine CB channels with their senseless chatter.

In 1986 fifty Vietnamese men worked twenty-four hours a day for eight days to build a community hall in Bethany Park. It had a large concrete foundation, and its roof and sides were sheet metal; inside it was roughly but adequately furnished. Every Wednesday night the Vietnamese priest celebrated mass for the community except for the men who were always out fishing. On Sundays the congregation still used the Catholic church in town because the new hall was not large enough to hold everyone when the men returned from fishing.

Catholicism was an integral part of the life and work of the Vietnamese fishermen, to a much greater degree than it was for the local Catholics. Since the Vietnamese fisherman could attend mass only on Sunday, they expressed their faith daily while working on the bay. Each morning at 8:00 and each evening at 5:00 one Vietnamese fisherman was selected to read the scriptures in Vietnamese to his fellow shrimpers; in this way they gained the religious strength that allowed them to continue the arduous work of fishing. The sounds that Texas shrimpers interpreted as Vietnamese gibberish tying up their marine CBs were in fact the reading of Vietnamese fishermen as they practiced their faith on the waters of Tres Palacios Bay.

Several different churches in Palacios made various attempts to proselytize the Vietnamese Catholics. Such efforts, according to the priest, were deeply insulting to the fishermen and their families. He suggested that the Vietnamese looked upon the attempts as callous, reflecting a gross miscalculation of the importance of the refugees' Catholic faith as an integral part of their everyday lives.

Even as the Vietnamese faced discrimination from the residents of Palacios, they also had serious problems within their own community, just as many other immigrants before them. Chief among them was the dilemma that the Vietnamese fishermen referred to as the "half generation," the young men who came with their families to the United States when they were in their late teens and twenties. The young single men found it difficult to adapt to the new cultures. The children in contrast learned English quickly and were more familiar with the American customs, and the older men and women clung to the traditional ways.

The half generation had unrealistic expectations of America, and in general they made bad fishermen. They were unwilling to work the long hours and wasted their money on unnecessary American products. They dressed differently from the other Vietnamese men, often wearing T-shirts and jeans judged inappropriate by the elders. They cut their hair differently, preferring the punk style common in larger cities. In Palacios the young men were quickly identifiable by their dress and grooming; they were not truly Vietnamese, according to the refugees in Bethany Park, nor were they Vietnamese-Americans. Sometimes called "teddy boys," the young men turned to petty crimes, including theft, and they were blamed for burning the van of one Vietnamese family. The family did not choose to report the crime to the Palacios police even though they knew who had set the van afire, believing that the police would not be able to solve the problem and that the teddy boys would seek revenge for being reported.

The teddy boys were not the only problem the Vietnamese faced from within. Normally infants and small children were looked after by the entire community, but in Bethany Park from the early morning until sunset there were few adults around. The majority of Vietnamese women, both young and old, worked at the crab plant, and almost all the men were out fishing or repairing their boats at the dock. The older children, except during the summer months, attended school. Proper care for infants and preschool children was therefore a growing concern.

The elders of the Vietnamese community were also worried that they were losing their culture and traditions in their new homeland. American-born children in Bethany Park learned how to speak Vietnamese within their families but not how to write it; indeed the majority of the adults, poorly educated peasants, did not know how to write their own language. Then in 1989 the public school system hired one Vietnamese teacher, and the priest started language classes in the trailer next to the new community hall.

Social myths about the Vietnamese perpetuated in Palacios and along the Texas coast portrayed them as extremely successful fishermen who were getting rich even as American fishermen were failing. In fact, Vietnamese fishermen in Bethany Park, just as other bay fishermen, were hard hit by falling prices in the mid-1980s; four families in Bethany Park had lost their boats by 1988. In each case, the families had gone to the local bank for a loan to help purchase their boat. Contrary to the social myth, which suggested that the

Vietnamese always paid for their shrimp boats with cash or with the gold they hoarded, these families, and many others in Bethany Park, borrowed money to build or buy their boats. When they had not been able to keep up the monthly loan payments, the bank repossessed their boats. The four families left Bethany Park to find employment in Houston, Dallas, or some other metropolitan area where jobs were more plentiful.

One Vietnamese girl whom I interviewed in Palacios was particularly incensed about the rumors that the refugees had become rich through welfare fraud. She told me that she and her mother, father, and six brothers and sisters had lived outside Houston before they were resettled in Palacios in 1976. Their foster family had been paid $300 to buy the Vietnamese family furniture for their small house; her family had been given only a set of bunk beds and no other furniture. The foster family had pocketed the difference. On Sundays, because the refugees did not have a car, they walked three miles to the nearest Catholic church to attend mass. As soon as they heard through the Vietnamese grapevine that there was room for them in Palacios they moved from Houston and never again talked to their foster family. She was still adamant that the foster family had cheated them.

In 1987 a small group of Palacios refugees went to Rome and had a brief audience with the pope. They had saved money for years to make the special journey, which they considered "the trip of a lifetime." It was the only time, with the exception of their flight from Vietnam, that any of them had ever been on a plane. The group was composed of several of the elders of the community, and their picture appeared in the *Palacios Beacon* shortly after their return from the Vatican. Some residents of Palacios interpreted the trip to Rome as yet another example of the affluence of the Vietnamese community.

Vietnamese shrimpers were motivated to work hard by the same drive that fueled the Texas shrimpers. They worked not just because they were used to it or because it was a part of their cultural tradition but because they had to pay off their debts. They owed money not only on their boats but also on their new lots and homes and on other possessions they valued, such as their pickups and vans. Three years after construction began in Bethany Park, about one-half of the lots had houses on them; the same trailers and used mobile homes that had been used in Bay Vue still sat on many of the new lots.

Vietnamese fishermen and their wives were not immune to the stress brought on by their labor and living conditions. I asked the priest to name their biggest problem aside from their troubles with the non-Vietnamese residents of Palacios. "The men work too hard," he told me. "They must always go fishing. The prices this year are not very good so they have to work more. Their wives are tired by the end of the day, always on their feet picking crab. Sometimes the families do not get along. They argue with each other. It's not a good situation. And who is looking after their children?"

Leisure time was at a premium. When the men returned from fishing trips, they maintained and repaired their boats. After the Vietnamese first came to Palacios, they spent much time repairing the old boats they had purchased in order to keep them in running condition. The worst were gradually replaced, but still much time was given to maintenance and repair. Constant duties and chores left the fishermen and their families with little time to think about applying for their papers to become American citizens, a lengthy and time-consuming process at best. When the fishermen found out it was necessary to be an American citizen to obtain a commercial operator's license to run a fishing vessel greater than five tons, some of them quickly rearranged their priorities. By 1989 about twenty men in Bethany Park had become U.S. citizens and could qualify to be captains of offshore Gulf shrimp boats, once they had passed a simple test and paid a small fee.

Some of the children of the Vietnamese fishermen and crab pickers found quick success in the public schools, just as they had in Seadrift. Many non-Vietnamese counted the achievements as an important measure of the immigrants' assimilation into the larger American culture, in their view a melting pot of racial and ethnic groups. Success, however, was not easily attained and, according to several Palacios respondents, was not without setbacks. One Vietnamese girl said that when she arrived in Palacios at the age of seven she knew no English at all but was placed in a regular second-grade public school class. Because no attempt was made by her teachers to help her learn English, she spent much of her first year in the new American school sleeping at her desk. Gradually she did learn some English from her Vietnamese friends and from television and became a good student who earned high grades. On the surface many of the Vietnamese children appeared to have adapted rather quickly to the Palacios public schools. Several Vietnamese boys had joined the high school basketball team, and one had joined the football team and was reputed to be above average in both ability and drive. At Bethany Park, next to the volleyball court and not far from the doors of the chapel, a wooden basketball goal with an iron hoop had been constructed. Vietnamese girls and boys joined the high school choir and participated regularly in other after-school activities, indicating their acceptance, at least at a certain level, by the other students.[27]

Five Vietnamese students from Bethany Park attended college. Males were encouraged to further their education, and college scholarships and loans were supplemented by monies from the Vietnamese community. In contrast, Vietnamese females were actively discouraged from going to college and were not given financial support for it. Instead, their families and others in Bethany Park encouraged them to marry shortly after graduation from high school. Several of the young girls rebelled against their parents' wishes and the Vietnamese traditions, which they identified as gender biased. One Vietnamese girl "ran away" to a nearby city where she lived with relatives while attending a community college. She said, "There's nothing to do [in Palacios] if you are an educated female. Nothing except the crab plant. There's no future here."

The relationship between the Catholic churches in Palacios and the Vietnamese community was strained. The town's Catholic churches were perceived by the Vietnamese to have been of limited help to them when they first arrived, providing little financial, emotional, or religious support for the refugees. Assistance from the Palacios Catholic community and from the nearest diocesan headquarters, located in San Antonio, was begrudging at best.

The Vietnamese fishermen always kept to themselves on the docks. When they shopped in Palacios for food and other necessities, they arrived and stayed in small groups. One family in Bethany Park opened a store specializing in Vietnamese foods, and another started a marine supply shop; the Vietnamese then had even fewer reasons to go into town to shop. Their boats were parked together, rafted up, separate from the boats of other bay shrimpers. The segregation was not of their choosing; Vietnamese were not allowed to dock at many Texas ports or were relegated to certain specific dock spaces, frequently at some distance from the rest of the boat traffic.

More than a decade after the refugees had settled in Palacios, the language barrier was still an important problem. Their boats were always crewed by other Vietnamese; it was extremely rare for a Vietnamese captain to hire non-Vietnamese or for a Vietnamese fisherman to work as crew on an American boat. The majority of the fishermen understood some English, often much more than they cared to admit, according to Vietnamese respondents. But their ability to fully understand English and, of equal importance, to use it to communicate with white residents of Palacios, including other bay fishermen, was limited.

To make matters worse, the monolingual Americans had little tolerance for the errors the Vietnamese made when speaking English. The white residents of Palacios saw the immigrants' lack of skills in English as another example of their failure to assimilate into American culture. Whites observed, with some accuracy, that the fishermen and their families were selective in their understanding and use of English; some Vietnamese obviously understood it when it was to their advantage. On the other hand, most white respondents failed to grasp that English was a difficult language for the newcomers to learn. Among Texans, however, a strong belief persisted that Vietnamese was an impossible language to learn; no Palacios residents that I met spoke even rudimentary Vietnamese. Except on two occasions, I never heard white residents mention any appreciation of or sensitivity to the culture and traditions of the refugees.

The Vietnamese community made several attempts to participate in the life of the larger community, to be good neighbors and citizens. For example, Bethany Park sponsored a float in the annual parade through the streets of the small downtown. One respondent told me that many of the children and adults spent long hours working on the float; they won first place for their efforts.

The bay shrimpers in the random sample survey were asked a series of four questions about their attitudes toward Vietnamese shrimpers.[28] To the first question, had they personally had any problems with Vietnamese shrimpers, a majority (76.3 percent) said they had not had any on an individual basis.

The problems they recounted centered on three different kinds of issues. There were frequent complaints that Vietnamese shrimpers did not regularly follow the same fishing practices, formal and informal, that American bay shrimpers followed. One bay shrimper said, "They drag in front of me. They run in packs like wolves." Another said, "They crowd you. They've cut me off and run over my nets." Another reported, "They come and drag all around you so you can't turn." Still another shrimper concluded, "They work around the clock because the whole family is on the boat. They eat anything they catch"; this respondent directly blamed the Vietnamese for a decline in shrimp and other species, including crab, in the bays of Texas.

A minority of bay shrimpers (23.6 percent) accused the Vietnamese fishermen of regularly breaking rules that governed fishing in the bay. One shrimper claimed, "I've seen them pull double rigs in the bay." Another said, "In San Antonio Bay the Vietnamese are dragging twenty-four hours a day. They have a bad attitude." One interviewee was still angry that the previous year several Vietnamese crabbers had accused him of stealing their pots. He found the charge against him particularly ironic because he felt that the Vietnamese were guilty of the same offense.

Texas bay shrimpers recounted situations in which the Vietnamese threatened them while fishing, including shouting at them, cursing, and displaying handguns and semiautomatic rifles; one shrimper characterized it as "showing me his gun." In one incident, a shrimper said, "I was trying to pick up. He came at me at a 90 degree angle. He stepped in his cabin and came out with a handgun." Sometimes the threats turned directly to violence. One Texas fisherman said his boat was forced against a sandbar by a Vietnamese boat. The Texan did not know why the Vietnamese shrimper did it, but "I had to shoot off a shotgun at 'em." Another recounted, "They hit the boat. Rammed it."

Old grudges died hard. One bay shrimper, even though eight years had passed, was still enraged, remembering a night when he was visited in his home by Vietnamese fishermen several weeks after he had spoken at an anti-Vietnamese rally in Kema. He had told shrimpers at the rally that "you ought to burn the Vietnamese boats like they did in Louisiana." Two weeks later a Vietnamese boat was in fact burned, and the Vietnamese fishermen who owned it came looking for him. The shrimper said he was innocent of any wrongdoing, but because he was afraid of them he moved away. Later, he charged that they "ran into my boat while docking. [Now] I stay away from them."

Such incidents suggest that many of the hostilities and much of the violence that occurred were not reported or documented. They also suggest that the conflict between the two groups was more widespread and serious than previously believed. Although shrimpers agreed that tensions had eased considerably since the early 1980s, they said the threat of violence was still a concern. When asked, "Are Vietnamese good bay shrimpers?" surprisingly, 75 percent of the fishermen answered yes. The American shrimpers generally admired the hard work of their competitors. "They just grind and grind," said one. Another said, "They're more organized than American fishermen are." One bay fisherman summed it up by saying, "They stay with it and that's what you have to do. That's all they've done all their life. They work. That's the problem. They're damn good fishermen."

Texas bay fishermen also attributed the perceived success of Vietnamese fishermen to their fishing strategies and their equipment. Many respondents answered that the Vietnamese were successful because they fished in groups and used "scouts." "These scouts, if they find the shrimp, then they call their friends and they all catch shrimp," reported one interviewee. Several shrimpers remarked that the Vietnamese do not use a try net, that they just drag their nets day and night, working longer hours than American shrimpers. Another admired the quality of their fishing nets; "They design their nets good and they aren't stupid."

The admiration was not without criticism. Texas shrimpers repeatedly accused the Vietnamese of not following the regulations governing fishing in the bay. Many blamed the Texas Parks and Wildlife Department for treating them with kid gloves. One shrimper summarized a common viewpoint: "They're good fishermen but they break the law and observe no limit." Another shrimper, explaining why he believed the Vietnamese caught more shrimp, said, "They clock [work twenty-four hours a day]." Several other reasons were given for their success, including the charge that "the main reason they succeed is that they put five to seven people on one boat." One shrimper said, "they were just born to it."

In every case, bay shrimpers believed that the Vietnamese were getting rich. They were not viewed as individuals, some as better fishermen, some as worse, as Americans viewed themselves. American shrimpers frequently made broad statements that they considered to be true for all Vietnamese fishermen. They saw the Vietnamese as a homogenous work force that labored beyond normal human endurance. There were no failures among them; they were big winners. The idea that there might be possible repercussions among Vietnamese fishermen because of their hardwork, that they and their families might suffer from the demands of their hard labor, was never given serious consideration.

Moreover, while the Texas bay shrimpers were quick to point out that the Vietnamese fishermen routinely broke limit restrictions on the size and poundage of the catch as well as rules governing seasonality and the usage of gear, they were equally quick to deny their own shortcomings. American fishermen admitted that they were "forced" to break rules regularly; indeed a good many of their fishing strategies were predicated on the breaking of rules governing the bay. But they argued that the Vietnamese "went too far." They criticized the TPWD for not fining or arresting the Vietnamese more frequently and complained about their own problems with the agency. In other words, they wanted the TPWD to fine the Vietnamese but to ignore their own lawbreaking. Texas bay shrimpers saw no inconsistencies in this opinion, nor did they understand why others who were not shrimpers might question the logic of their position.

The respondents in the sample were asked a final question, "Do you personally know any Vietnamese fishermen?" Surprisingly, 45.9 percent said that they did know at least one at a level they defined as "personal." The relatively high rate of response may help to explain why, almost ten years after the death of Billy Joe Aplin, the majority of Texas shrimpers had begrudgingly accepted the refugees. In the long run, the knowledge that came from personal contact with the Vietnamese may have mitigated against stronger negative attitudes toward them. A minority of fishermen, about one quarter of the sample, were still quite displeased with the presence of the Vietnamese. The majority of Texas bay shrimpers, however, had reached some level of acceptance, however qualified, a handful naming Vietnamese among their friends.

Fish house owners who directly served the Vietnamese had useful insights into the conflict between the two groups of fishermen on the bay, and those in Palacios were particularly perceptive about relationships between Texans and Vietnamese. Both owners whom I interviewed remarked on the Vietnamese as hard workers and were also sensitive to the costs of their hard work. One fish house owner was impressed with their integrity. "If they say they're going to do something, they do it. If they give their word, then you can count on it. That means a lot to me in this business. They are very reliable." He saw the problems between Vietnamese and Texas bay shrimpers in the context of an industry that was beset by a variety of constraints that hurt everyone, Vietnamese and Americans alike. He was also aware that several Vietnamese had failed to meet their loan obligations to the bank and were forced to forfeit their boats; he was the only non-Vietnamese I encountered who knew about this.

Three dockhands, all Texans, openly admired the Vietnamese fishermen. One had shrimped before, the other two were content to remain onshore even though they earned just above minimum wage. "Hey, look at it this way," one of them said, "they started with nothing. Now they have big trailers and even some houses. Give'em some credit."

Vietnamese bay fishermen and their families met with stiff resistance in Texas coastal communities.[29] The refugees in Port Isabel, Seadrift, and Palacios were scapegoated, blamed for a variety of activities over which they had little or no control. Hindsight, always perfect, would suggest that the timing of the resettlement of the Vietnamese along the Texas coast could not have been worse. Racism was rampant among local fishermen, and the Texans who did not express it were often sympathetic with those who did. Nevertheless, to the credit of many fishermen and other coastal residents, and despite the dominant media stereotyping, the Ku Klux Klan was unable to organize the racism or to push the individuals who fostered it farther to the political right.

In balance, however, the actions of the Vietnamese fishermen and their families in Port Isabel, Seadrift, and Palacios were not without fault. Whenever possible, they used their status as refugees to their own best advantage. They were consistently insensitive to the traditional fishing customs of the Texas bay shrimpers, and they were virtually inflexible in practicing their own fishing customs. The refugees were often unwilling to negotiate or to compromise on community issues in which they had a stake, pleading ignorance of English. In short, they were not the passive, naive, and innocent

Gulf shrimp boats line the harbor at Freeport, Texas.

victims of coldhearted racists. The media's stereotyping of the Vietnamese deprived them of their humanity in some of the same ways that it stripped the bay shrimpers of theirs.

The Vietnamese drove Texas bay and Gulf shrimpers together and gave them a common enemy to blame for their misfortunes. The real problem in the bay shrimp industry in the 1970s and 1980s was not the Vietnamese, but many bay shrimpers acted as if economic success was impossible unless the immigrants quit the shrimp fishery. In some coastal communities the racial hatred was worse than in others, but wherever Texas bay shrimpers and Vietnamese fished, the refugees were accused of wreaking havoc on the industry.

9

Losers and Winners

❖ ❖ ❖ ❖ ❖ ❖ ❖ ❖ ❖ ❖ ❖ ❖ ❖ ❖

Texas bay shrimpers are losing the battle against Texas Gulf shrimpers, against environmental groups, against state and federal government. Yet the battles are secondary to the war they have already lost. Prices for imported shrimp have battered domestic prices, making it hard for Texas shrimpers to earn a living from the bay. The world has changed, and the fishermen, like many other American workers, are feeling the effects of new economic circumstances.

Even in the best of economic times, shrimping Texas bays is demanding work that can wear out the toughest individuals. It is a mistake to romanticize the work of the commercial fishermen; theirs is a numbing routine that repeats itself every time they leave the dock. Bay shrimpers have found ways to contend with their work on the water; over the years, it has shaped every facet of their lives. Yet despite how hard they work at their job, they can never hope to sell their shrimp at a cheaper price than South American or Indonesian peasants can grow them. Shrimp raised in ponds and bays capitalized by foreign governments and private investors are flooding the American market, driving more and more Texas bay shrimpers out of business.

Although production in Ecuador and China fell in 1994, countries such as Indonesia appear poised to increase their exports to

A bay captain carefully boards his net after fishing for an hour and a half.

American shores. Domestic shrimp prices, barring some unforeseen increase in demand, will continue to stagnate, driving all but the most successful and resilient bay shrimpers from their work. Texas bay shrimpers will survive, but they will be far fewer in number, make lower profits from their fishing, and rely increasingly on off-boat jobs to remain solvent. Local fresh shrimp markets, which have supported a bay industry since the 1870s, will not disappear, but shrimpers who supply them have taken a beating from which they will not recover.

Furthermore, the battle over Texas bays has not ended; bay shrimpers may yet be driven completely out of business. Environmental groups have been successful in getting their way with TEDs. BEDs are potentially the last straw for them; if not BEDs, then some other environmental issue may well emerge. Bay shrimpers still use fishing gear and techniques that have not changed significantly since the 1930s, but environmental concerns, or at least needs and wants placed under that mantle, have changed. Environmentalists now have the law, the money, and the political clout to force bay shrimpers to stop fishing. The real question is whether environmentalists will stop short of eliminating the industry so that both private and commercial use of the Texas bay fishery can continue.

It is difficult to be optimistic about the future of bay shrimping in Texas because of the Gulf Coast Conservation Association. The GCCA, a broad-based organization with a serious environmental agenda, has amassed a wide membership throughout Texas that gives it both lobbying dollars and enormous political power. From its earliest beginnings it has been organized to reflect the particular desires of a small group of wealthy Texans. With its influence over the Texas Parks and Wildlife Department, the GCCA controls the destiny of the bay shrimp industry.

Fishery managers in Texas have been placed in an uncomfortable position not of their own making. The money, the political power, and the law are on one side; on the other are shrimpers trying to make a living from the bay. The majority of managers have shifted their allegiance to the controlling powers, but a minority have been more vocal, pointing out inconsistencies in the established policies and in the laws and regulations. Most Texas fishery managers have ended up as apologists for the status quo, frontline bureaucrats who must contend with angry bay fishermen who question the laws, doubt the social science used to justify them, and deny the integrity of the decisionmaking process. Fishery managers are buffers between the people who make the law and the people who suffer its enforcement. They also serve as policy rationalizers, legitimizing political decisions that undermine the bay shrimp industry. In turn, the bay shrimpers blame the managers for a variety of problems the managers had nothing to do with. It is unfortunate that many of the managers are so far removed from the daily lives of bay shrimpers and their work; they have made few attempts to learn about the fishermen, to familiarize themselves with the shrimpers' work on the bay, their families, and their fishing communities.

A crude stereotype of bay shrimpers pervades most discussions of environmental and policy issues. They are commonly characterized as loud, stupid, dirty, rude, and greedy, coastal hicks unable to understand the benefits of whatever new policy or law is being written. In recent years bay shrimpers have been redefined as poachers who destroy marine resources and rob the public. The realities that belie the stereotype are quite complex, however. Bay shrimpers are small-scale entrepreneurs who dress in jeans and T-shirts because of the nature of their work. They own their boats and historically have earned substantial profits from harvesting shrimp; their industry has sustained their coastal communities economically. Sometimes they are loud and make rude comments, their frustration and anger particularly directed at people unsympathetic to their problems. Although they live in rural fishing communities, they are certainly not hicks; their business skills may equal or exceed the skills of the bureaucrats and lawmakers who have been delegated to manage their industry.

Bay shrimpers are thus forced into a no-win situation. Because they are trying to stay in business, they break the laws in the 1990s, laws that have been developed and enacted by their competitors and political enemies. At the same time, their vilification has been insidious; they are called poachers. In a classic catch-22 situation, the fishermen develop innovative ways to circumvent the use of TEDs because they know from their own experience that the devices, despite the promises of governmental agencies to the contrary, reduce their production. Yet since they are fined if they tamper with the TED, in their view they must catch more shrimp to make up for the loss.

There are only a limited number of ways to circumvent fishing restrictions; if enforcement is increased, many bay shrimpers will have to quit the business. Because family resources can be stretched only so far, shrimpers' wives will continue to play an important part in the family business by working as deckhands and by accomplishing other necessary shore tasks. The additional work, however, takes its toll on the women and their families. More than one wife described her exhaustion from working as a deckhand and raising children. Whether they work with their husbands or at other jobs, shrimpers' wives are cursed by the demands of the second shift.

Bay shrimpers must catch more and more shrimp to stay in business at a time when there are too many fishermen and not enough shrimp. Currently the Texas Parks and Wildlife Department has mandated a form of limited entry upon bay shrimpers; a freeze on bay boat licenses is in effect while specific limited entry policies are developed. The TPWD held a series of meetings along the Texas coast in 1994 nominally to collect information from bay shrimpers about their opinions on what kind of limited entry they preferred; there are a variety of different options available. Participation from bay shrimpers in these public meetings was uneven; many fishermen neither want limited entry, in any form, nor do they trust the TPWD to represent their best interests. Many Vietnamese bay shrimpers are also distrustful of any policy that would change the status quo. It is impossible at this time to gauge the long-term impact of an as yet unspecified limited entry policy upon bay shrimpers and their industry; however, one can predict with some degree of certainty that the limited entry policy, whatever its final form, will directly benefit both the Texas Gulf shrimpers and the Gulf Coast Conservation Association.

It is understandable that bay shrimpers, lost in a bewildering maze of unsatisfactory choices and decisions, have periodically blamed various players involved in the industry. Fishery managers have shouldered their share of criticism along with the media and other groups and institutions. The shrimpers' frustration and anger also fell on Vietnamese refugees who were resettled in coastal communities in the 1970s. It is not the first time that immigrants to this country have been blamed for problems they did not create. But the events surrounding the killing of Billy Jo Aplin, despite the media blitz, did not reaffirm the stereotype of bay shrimpers as racist rednecks; rather, their reaction to the incident was clear proof that the fishermen are much more complex than the stereotype would suggest.

The resettlement of Vietnamese along the Texas coast is an excellent example of American immigrants surpassing seemingly insurmountable odds. Through their hard work, Vietnamese bay shrimpers have attained material success in a few short years; yet it would be a mistake to ignore the costs of their success in terms of their families' lives. Their assimilation into the American melting pot is also misleading; it remains to be seen if the second and third generations will face similar struggles against racism.

Bay and Gulf shrimpers are still at war over the shrimp fishery, as they have been for almost fifty years. The Gulf shrimpers persistently believe that if they could put their bay competitors out of business, their profits would soar. Gulf boat owners, according to stereotype, are supposed to be more astute at business because their revenues are much higher than those of their bay counterparts. Yet their understanding of the politics and economics of the shrimp industry can be just as narrow as the fishermen who captain bay boats. Both bay and Gulf fishermen and boat owners face similar kinds of problems in the global shrimp market. Their infighting helps neither side; ultimately bay and Gulf shrimpers have many more reasons to band together to solve their mutual problems than to continue their hostilities.

As bay shrimpers drop out of commercial fishing, they will be hard put to find jobs that match their income from commercial fishing. Younger men and their families, moving to better job markets in the cities, may find it less difficult to walk away from the shrimping life. Older fishermen probably will stick with their lifelong occupation until their creditors close them down, until they can no longer purchase gas and ice for their boats and make major repairs. They will choose their familiar homes over moving away, finding minimum-wage jobs and making do as best they can.

Local economies in Texas fishing communities will suffer in direct proportion to the number of bay fishermen who lose their businesses; bay fishing is an integral part of the economic life of these small towns. No new industries will rush to replace the old; in the absence of any planned development projects funded by state or federal dollars, the coastal towns will suffer. State-supported programs are necessary to retrain bay shrimpers and to bring new industry to their communities.

Bay fishing is also a vital part of the cultural identity of these small towns, supplying rich family and community traditions that delineate standards for the individual in society and that detail personal and social values and expectations. Shrimping is the measure by which one is judged, known, and respected in the community; it is not just a job but a proscription for how to lead one's life, defining who one was, is, and will be.

The economic stagnation and loss of social identity in the Texas coastal communities is admittedly a relative issue when compared to the problems faced by thousands of other small towns throughout the United States. But Americans face a unique loss if we lose Texas bay shrimpers and their industry; their most important value to society may well be much greater than their annual profits from fishing the bays. They can tell us what is happening in our bays and estuarine systems based on their knowledge of the waters and their long-term stake in them.

The weekend angler and even the big boat crowd spend little time on the bay; frequently their boats are tied up at the dock. They cannot know the bays intimately. When the waters become polluted or other environmental issues are raised, recreational and sports fishermen trailer their boats somewhere else. It is a mistake to believe that inshore and offshore hobby fishermen can thoroughly become knowledgeable about their coastal environment and capable of reporting changes in its status. Though environmentalists per se have a greater commitment, they can also pick the wrong target for their efforts, as the case of bay shrimpers suggests. And all too often scientists and technicians working with government and business have a vested interest in receiving grants and advancing their careers.

Bay shrimpers have never been stewards of the bay, certainly not in the way that some small farmers have been stewards of their land. But what has been good for the bays ultimately is good for the shrimpers and their families. They know the bays, estuaries, rivers, and beaches better than anyone else. Their knowledge comes from the time they spend and have spent within this social and biological environment. Their well being is determined by their knowledge of the region. It has always been in their best interests for the bays to be cared for, protected from industrial pollution, toxic waste disposal, coastal development, and overfishing.

Texas bay shrimpers face a number of major challenges in this decade. Often complex in nature, they are closely tied to fundamental changes in the structure of our world economy and polity. A wide variety of Americans face similar challenges: even though it is a long way from the work deck of a bay shrimp boat to the offices of IBM or the factory floor of General Motors, American workers are losing their jobs in record numbers to foreign competition. And new jobs often pay minimum wage or little more, offer few benefits, and provide limited possibilities for advancement.

Texas bay shrimpers clearly have felt the impact from the globalization of capital within our national boundaries. Yet their situation indicates that political and economic forces are neither necessarily predetermined nor immutable and that they involve the active participation of various levels of government, business, and other social institutions. Although Texas bay shrimpers see their problems from their own perspective, we must do otherwise. We must understand and appreciate the commonalities between the fishermen and other American workers who, through little fault of their own, are losing their jobs and their way of life.

NOTES

❖ ❖ ❖

Introduction

1. See "Fisheries of the United States 1990," National Marine Fisheries Service, National Oceanic and Atmospheric Administration (NOAA), U.S. Department of Commerce (USDC), (Washington, D.C.: GPO, 1991).
2. Robert Lee Maril, *Texas Shrimpers: Community, Capitalism, and the Sea* (College Station: Texas A & M University Press, 1983).
3. See Norman K. Denzin, *The Research Act: A Theoretical Introduction to Sociological Methods*. (Englewood Cliffs, N.J.: Prentice-Hall, 1989).
4. There are 142 captains in the random sample survey and twelve deckhands. The survey was collected during two fishing seasons in which catches were relatively low. A majority of captains had stopped using deckhands in order to save on overhead costs, as was normally the case. Data reported about bay fishermen, unless otherwise noted, include both captains and deckhands. If I were to repeat this study, I would collect a random sample of the bay boat captains and a purposive sample of deckhands. Some of the problems confronting a researcher who intends to collect a random sample of shrimpers, as well as the value of using participant-observation to study these fishermen, are discussed in detail in Robert Lee Maril, "Methodological Approaches to the Study of American Commercial Fisherman," *Proceedings of the Workshop on Fisheries Sociology*, (Technical Report) Connor Bailey et al., eds., Woods Hole Oceanographic Institute, Woods Hole, Mass., 1986.

CHAPTER 1
The Men, the Boats, and the Fish

1. See Donald W. Field et al., *Coastal Wetlands of the United States: An Accounting of a Valuable National Resource* (Rockville, Md.: NOAA, 1991).

See also *Estuaries of the United States: Vital Statistics of a National Resource Base* (Rockville, Md.: NOAA, 1990).

2. Data from an Environmental Protection Agency (EPA) study as cited in Linda Kanamine, "Report: Toxic Emissions Down but They Still Pose a Health Risk," *USA Today*, October 1, 1992, 13A.

3. See *Estuaries*, and Anthony J. Reyer et al., *The Distribution and Areal Extent of Coastal Wetlands in Estuaries of the Gulf of Mexico* (Rockville, Md.: NOAA, 1988).

4. See, for example, NOAA/EPA Team on Near Coastal Waters, "Susceptibility and Status of Gulf of Mexico Estuaries to Nutrient Discharges," Summary Report, Washington, D.C., June 1989.

5. See Robert Lee Maril, *Cannibals and Condos: Texans and Texas along the Gulf Coast* (College Station: Texas A & M University Press, 1986), and Martin Salinas, *Indians of the Rio Grande Delta* (Austin: University of Texas Press, 1990). See also Thomas R. Hester, "A Chronological Overview of Prehistoric Southern and South-Central Texas," in *Prehistory of Northeastern Mexico and Adjacent Texas*," ed. Jonathan Epstein, Thomas R. Hester, and Herbert Graves (San Antonio: Center for Archaeological Research, University of Texas, 1980), 23–26.

6. The random sample of 154 bay shrimpers included only men, but after I interviewed male shrimpers both in and outside the sample, it became clear that women play an important role in fishing for shrimp in Texas bays, in contrast to the Texas Gulf shrimp industry (see chapter 7).

7. The sample size of 154 bay shrimpers represents 1.8 percent of the estimated total number of Texas bay shrimpers. At the time of the survey, no reliable data recorded their number; in the past, estimates have varied dramatically, depending on the methods used. My estimate of the number of Texas bay shrimpers is 5,453 fishermen, based upon the number of shrimp bay licenses sold during 1988 and 1989, the years of my interviews. The average number of bay licenses for the time period was 2,908; I multiplied this number by a factor of 1.875 to include the captain and the average number of deckhands on each bay boat. In my survey I found that 87.5 percent of the captains said they regularly used one deckhand all the time. The estimate remains biased in several ways, but I believe it to be the best, given the limitations of the existing information about the fishermen.

8. Several problems arose in collecting data on the different racial and ethnic groups of bay shrimpers. I found it difficult to include Vietnamese fishermen in the survey. First, their recent troubles and the publicity surrounding them resulted in their not wanting to be interviewed. Second, the regulatory issues affecting shrimp fishing in the

bays confused and angered them so that they were unwilling to be
questioned. Third, they suspected me because they believed I was a social scientist working for people whom they did not trust, despite my
denials to the contrary. Although I made every effort to interview them
over a period of three months, I finally was forced to rely on the information of Vietnamese informants who were familiar with the fishermen and upon my study of Alabama Vietnamese fishermen, some of
whom had lived in Texas or were about to move there.

The Galveston Bay area, where there are a number of African-Ameri-
can bay fishermen, was intentionally excluded from the sampling frame
because a study was recently completed there by another research team.
After that study, the bay shrimpers had become even more disturbed by
state regulations governing their fishing; it was my judgment that reinter-
viewing these men would result in unreliable data. I have used and cited
data from the Galveston study when necessary.

Whites in the original sample of respondents composed 82 percent
of the people interviewed, Mexican Americans 15 percent, African
Americans 3 percent, Mexicans 1 percent; no Vietnamese were included. A corrected estimate was then established, based on the
known undercounts and on my experience in counting similar populations of Gulf shrimpers. I hope that future research will benefit from
the description of the problems I encountered in sampling this population.

9. These general statements are based upon the following data:

Age	Deckhand	Captain
16–19	0	2
20–24	7	4
25–29	3	16
30–34	0	12
35–39	2	17
40–44	0	21
45–49	0	21
50–54	0	19
55–59	0	15
60–64	0	11
65 or more	0	4
Total	12	142

10. These conclusions are based upon the following data:

Years of education	Deckhand	Captain
0–6	0	14
7–9	4	27
10–12	5	71
13–16	3	25
17 or more	0	1
Total	12	138

11. Based upon the 1979 survey sample of Texas Gulf shrimpers, it was estimated that whites composed 27 percent of the Texas labor force of Gulf fishermen; Mexican nationals, Mexican Americans, and blacks were at 20 percent each, Central Americans at 7 percent, and Cajuns at 6 percent. See Robert Lee Maril, *Texas Shrimpers* (College Station: Texas A & M University Press, 1983), 53–55.

12. The influence of tourists on bay fishermen in Corpus Christi was varied and considerable. Enterprising fishermen hauled to the docks what might best be called "symbolic" boats that did not have working engines or in some cases rudders but that were outfitted with a number of small coolers filled with ice. Shrimp were offloaded onto the boats, from which "fresh" shrimp were then sold to tourists.

13. Respondents on 150 different bay boats were sampled.

14. See Eugene A. Laurent, "Description of Industry: Economics," in *The Shrimp Fishery of the Southeastern United States: A Management Planning Profile,*" ed. D. R. Calder, Technical Report 5 (Charleston: South Carolina Marine Resources Center, 1974), 70–85.

15. See David R. M. White, "Environment, Technology, and Time-Use Patterns in the Gulf Coast Shrimp Fishery," in *Those Who Live from the Sea*, ed. M. E. Smith, American Ethnological Society (St. Paul, Minn.: West Publishing Company, 1977), 195–214. See also David R. M. White, "Skipper Effect or Fleet Effect? How Alabama Shrimpers Find Shrimp," paper presented at the meeting of the Society for Applied Anthropology, Charles, S.C., 1991.

16. The best example may be the large statue of a shrimp that stands in Aransas, Texas.

17. White, "Skipper Effect or Fleet Effect?"

18. Laurent, "Description of Industry."

CHAPTER 2
Working on the Water

1. The work arrangements on Texas bay shrimp boats differ from those on Texas Gulf shrimp boats. On the much larger offshore boats there is a captain, a rigman, and at least one header. Although there is some overlap in work duties and expectations, in general each position is clearly defined.

Like the captain of the bay boat, the captain on Texas Gulf boats is in charge of operations, maintenance, and repair; he also makes the important decisions regarding fishing strategies and is responsible for the men and the vessel, which he usually does not own. The rigman on a Gulf boat, as the name suggests, helps the captain handle the four large nets, stands watch at the wheel, heads the shrimp, and is the captain's right-hand man. The headers, the lowest position on board, spend the majority of their time deheading shrimp.

A header on a Texas Gulf trawler is an apprentice who may be promoted after several years to rigman if he shows promise. A rigman has to have from three to five years additional experience to be considered seriously for the position of captain. Rigmen are often given tryouts as captains, and if they do not prove satisfactory to the owners of the trawlers, they remain rigmen.

The annual incomes of Gulf shrimpers vary widely, depending on a number of factors, and reflect the differences in responsibilities of captain, rigman, and header. Headers earn the least, a piece-rate, but rigmen earn a share of the catch, which is two to three times a header's annual income. Captains earn a larger share of the catch than rigmen although the best of the rigmen earn more than the lowest-earning captains; still, the highest-earning captains earn twice or more the pay that the average rigman brings in annually.

When I first began studying Texas Gulf shrimpers, I assumed that Gulf boats from other states maintained the same set of work positions relationships, and incomes as that found on Texas Gulf trawlers, but my assumption proved false. Mississippi, Alabama, Louisiana, and Florida Gulf boats have deckhands, just as Texas bay shrimp boats do. The captains on trawlers from other Gulf states hire one or two deckhands or more, as required. They may have different skills and experience, but they engage in the same kinds of work, unlike the workers on Texas Gulf boats. In general, deckhands are paid the same share.

Why, then, are the positions on board Texas Gulf trawlers different from those of the Texas bay shrimp industry and the other Gulf states' shrimp industries? I posit both a historical and an economic explanation. The Texas bay industry predates the Texas Gulf shrimp industry by at least sixty years; the work relationships on board bay boats were established long before the Texas Gulf industry began. The latter developed rapidly after World War II and in the early 1950s; Texas was the last Gulf state to develop a significant offshore shrimp fishery. Florida and Louisiana Gulf boat owners, many of whom were Cajuns, migrated to Texas ports in order to be close to the Texas waters and the Bay of Campeche after the discovery of brown shrimp. The new labor force for these Texas Gulf trawlers was drawn from along the Texas Gulf Coast but particularly from the major fishing ports of Aransas Pass and Port Isabel–Port Brownsville. The populations of these and surrounding communities were largely impoverished Hispanics, with whites forming a more affluent minority (see Robert Lee Maril, *Poorest of Americans: The Mexican Americans of the Lower Rio Grande Valley of Texas* [Notre Dame, Ind.: University of Notre Dame Press, 1989]). These men made up a willing labor force, anxious for work, eager to learn new skills, especially hardworking, and soon highly motivated by the possibilities for advancement on a Texas shrimp trawler.

To save money and to increase profits, owners of the Texas Gulf boats replaced the generalist position of deckhand with that of a rigman in charge of one or more headers. Initially, the captain and rigmen were white; the headers, however, were almost always poor Mexican-American and Mexican men with no fishing experience.

The labor force of Mexican Americans and Mexican nationals was economically exploited by Texas Gulf shrimp boat owners; Mexican nationals were paid approximately one-half the wages of their American counterparts (see Robert Lee Maril, *Texas Shrimpers* [College Station: Texas A & M University Press, 1983], 16 and 31). They continued to earn less in 1990 than did American citizens employed in the same work; a recent program under the auspices of the U.S. Department of Labor has raised wages for Mexican nationals to approximate the level of American workers.

A number of Mexican Americans and Mexicans in Port Isabel and Port Brownsville worked their way up through the hierarchy by their labors on the rear decks of Texas shrimp trawlers. Although they began as headers, by the mid-1960s the Hispanics were hired as captains because of the hard work they had demonstrated, the expertise they had acquired, and their annual productivity. In a few cases the same captains were able to purchase Gulf trawlers themselves, and several of the most ambitious built up small fleets in the 1970s and 1980s. The tradition of hiring a rigman and headers rather than deckhands still prevails on Texas Gulf trawlers. (For a more detailed discussion of the role of captains, rigmen, and headers on Gulf boats, see Maril, *Texas Shrimpers*, 9–31.)

2. There are significantly more fatalities and injuries in commercial fishing in Alaska than elsewhere in the United States. As an occupation, commercial fishing is among the top ten most dangerous per 100,000 workers, preceded by mining and construction (see the study by the National Institute of Occupational Safety and Health on workplace deaths cited in "Taxi Drivers Face Most On-Job Murder," *USA Today*, November 29, 1993, 2A).

3. There are several excellent studies on job satisfaction among commercial fishermen. See, for instance, John B. Gatewood and Bonnie J. McCay, "Job Satisfaction and the Culture of Fishing: A Comparison of Six New Jersey Fisheries," *Maritime Anthropology Studies* 1 (1988): 103–28, and idem, "Comparison of Job Satisfaction in Six New Jersey Fisheries: Implications for Management," *Human Organization* 49, 1 (1990): 14–25. The findings of these and other studies are particularly sensitive to the regulatory problems that fishermen may face at the time they are interviewed; the fishermen's temporary problems may heavily influence their responses.

4. In 1985 in the federal courthouse at Brownsville, Texas, I reviewed the records of lawsuits brought by shrimpers against boat owners in Port Isabel, Port Brownsville, and Port Aransas; these fleets represent the largest along the Texas coast. Texas Gulf shrimpers injured themselves more frequently and more seriously in their work at sea than did bay shrimpers. Indeed in several Texas port cities there are small but thriving law firms or branches of larger Houston and Corpus Christi firms that specialize in the preparation of cases involving Texas Gulf shrimpers who sue boat owners.

The Brownsville court records showed that lawsuits among Texas Gulf shrimpers substantially increased during the early 1980s. There were thirty-one personal injury cases brought by Gulf shrimpers in 1980, forty in 1981, and forty-four in 1982. In 1983 the number of lawsuits almost doubled, to seventy-six, remaining at seventy-five in 1984. In 1985 the number of lawsuits was estimated at seventy-one. Lawsuits brought in state court cases in the region also reflected a significant increase during the same period, from an average of forty-five cases per year from 1978 through 1982 to sixty-four cases in 1983 (personal correspondence, Marine Advisory Bureau, Combes, Texas, 1986).

Although lawsuits increased from 1983 through 1985, there are no data that suggest a corresponding increase in injuries to Gulf shrimpers during this time. Moreover, there were no major storms or hurricanes and no changes in fishing gear, strategies or hitherto ignored flaws in vessel design or equipment that could account for a doubling in injuries. Lawsuits clearly increased among Gulf shrimpers during this period because of an aggressive legal community in South Texas that enlisted and actively recruited plaintiffs. Several of the situational components were already in place prior to the dramatic increase in lawsuits. Of the men who sued from 1978 to 1983, 44 percent did so in late summer and early fall; at this point a Gulf shrimper has earned most of his annual income from shrimping. In other words, Texas Gulf shrimpers worked through the most productive part of the fishing season when they earned the highest wages and then sued when the lean months approached. This observation does not discount the legitimacy of the fishermen's claims; some plaintiffs undoubtedly had suffered injuries. But the marked increase in lawsuits in the 1980s was not the result of an increase in injuries among Texas Gulf shrimpers.

Almost half (45 percent) of all lawsuits from 1978 to 1983 were brought by shrimpers between the ages of twenty and thirty; the vast majority of these workers were headers. About 80 percent of the cases involved injuries to the back, most often from falls on the rear work deck or in the hold.

The real wages of shrimpers in the early 1980s declined substantially in South Texas, partly because of the renewed enforcement by the U.S. Coast Guard of the Lacey Act, which prohibited Texas shrimpers from fishing in Mexican waters. The shrimpers complained that the Coast Guard's efforts cost them as much as 25 percent of their annual income or more, and the rise in the number of their lawsuits corresponds to the enforcement of the act. Meanwhile, other factors affecting the Texas Gulf shrimping industry came into play, including rising imports and falling prices, which also suppressed wages for the fishermen.

As more and more young shrimpers, particularly headers, faced a future of declining wages, the opportunity for turning both real and frivolous injuries into large cash settlements grew increasingly attractive. Word-of-mouth at the docks was pervasive. Some lawyers hired runners to bring them cases; as soon as an injured shrimper returned from his fishing trip, the runner communicated that news along with other important information about the fisherman to the lawyer. The traditional social and cultural barriers against suing Gulf boat owners, still recognized among bay shrimpers, began to fall.

Informants at Port Brownsville and Port Isabel said that among Texas

Gulf shrimpers it became common knowledge that the best complaint, the one that would bring the most money before a sympathetic local jury, was a back injury. Back injuries were difficult to disprove in court; sometimes they did not show up in X-rays, because the damage was to the soft muscle tissues. Nevertheless, the pain one felt from a legitimate back injury could be severe and debilitating. Some Texas headers discovered what kind of injury they should have and then conjured it up. They learned that by combining a painful back injury and a good lawyer they might hit the jackpot, getting $100,000 or more in cash, tax free. And if they lost in court, they could always go back to work as a shrimper.

A doctor's examination was of course a prerequisite; after that, if the results proved encouraging for the plaintiff's lawyer, a suit was filed. In 1985 only five Brownsville lawyers accounted for 83 percent of the shrimpers' injury cases filed in federal court; one lawyer alone filed 30 percent of the cases. This handful of local lawyers relied upon an even smaller number of medical doctors to examine their clients and, in the case of back injuries, to perform one or more corrective surgeries. The majority of operations were completed before a settlement or a trial and served as additional evidence that the plaintiff had suffered a serious injury.

Nineteen eighty-six was an unusual year for Gulf shrimpers. Production skyrocketed and prices remained relatively stable; shrimpers earned from one and a half to double or more of their usual annual wages. That year, according to interviews with lawyers, shrimpers, and others associated with the fishing industry, the number of lawsuits declined substantially. Shrimpers were earning excellent money and ignored their minor injuries; headers were far less motivated to invent injuries and to take them to court. But in 1987, a production year resembling those from 1980 to 1985, lawsuits rose again to their previous levels. The lure of hitting it big brought the headers and their lawyers back to court.

In response to the increases in the number of lawsuits, settlements out of court, and trials, insurance carriers began to drop their policies on Texas Gulf trawlers, and those companies that carried trawler policies raised their annual premiums. For example, hull and machine coverage on a trawler rose from an average of $8,700 in 1980 to $13,800 in 1985, while during those same years costs for protection and indemnity of the vessel rose from $3,461 to $7,368 (see U.S. International Trade Commission, *Conditions of Competition Affecting the U.S. Gulf and South Atlantic Shrimp Industry* [Washington, D.C.: GPO, 1985], 248). Rising premiums forced many Texas Gulf boat owners to drop their policies, and they became far less attractive as defendants to the shrimpers and their lawyers who might have considered a lawsuit against them.

A handful of boat owners in Port Isabel and Port Brownsville adopted an antisuit strategy in the late 1980s. They announced to their crews that workers who were injured would have their medical bills paid promptly, as always, but those who were injured and who decided to sue would have little to collect in a judgment in their favor because the owners would not be carrying insurance. At least on paper, the boat owners were cash and asset poor; by "going bare," they hoped to avoid frivolous lawsuits against them. By 1994 this strategy had been adopted by a growing number of boat owners.

A lone critic of the marine insurance industry offers another view of the rising number of legal suits against Texas Gulf boat owners. Neil C. Danque argues that the major cause of the increase in lawsuits among Gulf shrimpers stems from the way in which the insurance companies treat plaintiffs. Unlike automotive insurers, for example, the boat insurers do not employ claims adjusters; this practice forces the injured shrimper to bring suit in order to be recognized. Further, boat insurers, fearful of "monster" jury decisions against them, do not hesitate to settle out of court for substantial cash payments to shrimpers with questionable claims. These two practices, among others, contribute heavily to the rise in lawsuits (see Danque, "The Causes of the Liability Insurance Crisis in Commercial Fishing," Belleville, Tex., *May Day News*, n.d.).

By the 1990s insurance companies were no longer recognized as responsible for certain kinds of damages brought against them for a shrimper's wrongful death. In a Supreme Court decision for the companies, a shrimper's lost future earnings could no longer be considered part of the economic loss suffered by his survivors (Mercedel W. Miles v. Apex Marine Corporation, 498 US, 112 L. Ed. 2d 275, 111 S Ct., no. 89–1158). Consideration of a shrimper's future earnings on occasion had led to out-of-court settlements and jury verdicts of greater than $1 million; in turn, such large settlements had served to fuel additional lawsuits.

CHAPTER 3
The Political Economics of Bay Shrimping

1. The majority of bay captains (74.5 percent) split the sale of the shrimp with the owner on a fifty/fifty basis, and about 8 percent split it

thirty/seventy. About 6 percent split it seventy/thirty or forty/sixty; less than 4 percent split it twenty/eighty. Other arrangements were rare.

2. See Robert Lee Maril, *Texas Shrimpers* (College Station: Texas A & M University Press, 1983), 159–63.

3. In 1988 I conducted a study of fifteen oystermen in and around Bayou La Batre, Alabama. One of their biggest complaints was about the illegal oystering they witnessed on an annual basis. Yet the fishermen, some of whom were also shrimpers, stated that such oystering, illegal in light of new regulations, was necessary for them to make a living. They reported that illegal oystering occurred in Texas, Louisiana, Mississippi, and Alabama. My general impression from the interviews was that a substantial market for illegal oysters existed.

4. See Wade L. Griffin and Lonnie L. Jones, "Economic Impact of Commercial Shrimp Landings on the Economy of Texas," *Marine Fisheries Review* 37, 7 (1975): 13–14.

5. See Maril, *Texas Shrimpers*, 55–59.

6. This practice may have changed recently. When I interviewed Gulf shrimpers in Port Isabel and Port Brownsville in 1994, the season had been reduced to six months and catches were significantly lower. Such circumstances would make it much more likely for fishermen to seek other employment when they were not fishing.

7. Among other ways of determining wages is a custom known as "splitting the deck." Two or more men work the traditional jobs of header and rigman, equally dividing the wages they receive for their work. Occasionally headers make wage arrangements among themselves, unbeknownst to the captain. Two experienced men may fish a Gulf boat without any headers or rigman, splitting the boat's share of the earnings fifty/fifty.

8. I collected wage data for Gulf shrimpers in Port Isabel and Port Brownsville in 1979, 1983, 1986, 1989, and 1993; in 1989 and 1993 the data collection was funded by the Texas Employment Commission. The trip sheets that recorded the wages paid to a boat's crew were collected; boats averaged from seven to fourteen fishing trips per year. A mean average wage was calculated for the captain, rigman, and first header, along with an average box rate the header received. Between thirty and forty low-, medium-, and high-producing boats were sampled, including freezer boats that spent up to six weeks or more at sea.

9. See Texas Parks and Wildlife Department, *A Guide to Commercial Fishing Regulations*, PWD-BK-3400, 74-8/88 (Austin: Fisheries and Wildlife Division, 1988), 13–14.

10. Ibid., 13. The development of these laws and policies is closely examined in following chapters of this book.

11. There was also in the past a recreational license for noncommercial fishermen, which permitted fishermen to net up to fifty pounds of shrimp for their personal use; they were not allowed by law to sell it.

12. Texas shrimpers are not the only shrimpers who migrate; see Jeffrey C. Johnson and Michael K. Orbach, "Migratory Fishermen: A Case Study in Interjurisdictional Natural Resource Management," *Ocean and Shoreline Management* 13, 304 (1990): 231–52.

13. See David R. M. White, "Environment, Technology, and Time-Use Patterns in the Gulf Coast Shrimp Fishery," in *Those Who Live from the Sea*, ed. M. Estellie Smith, American Ethnological Society (St. Paul, Minn.: West Publishing Company, 1977), 195–214; idem, "Knocking 'em Dead: The Formation and Dispersal of Work Fleets among Alabama Shrimp Boats," *Maritime Anthropological Studies* 2, 1 (1989): 69–79; idem, "Skipper Effect or Fleet Effect? How Alabama Shrimpers Find Shrimp," paper presented at the annual meeting of the Society for Applied Anthropology, Charleston, S.C., 1991.

14. See, for example, E. Paul Durrenberger and Gisli Palsson, "To Dream of Fish: The Causes of Icelandic Skippers' Fishing Success," *Journal of Anthropological Research* 38 (1982): 227–42, and David R. M. White, "Skipper Effect or Fleet Effect?"

15. Vietnamese fishermen in one Texas coastal community I studied lost five boats during a two-year span; the boats were originally purchased by loans from a local bank.

16. The role of women in shrimping the bays is discussed in greater detail in chapter 7.

17. The trend has been documented in detail; see, for example, Peggy F. Barlett, "The Crisis in Family Farming: Who Will Survive?" in *Farmwork and Fieldwork*, ed. Michael Chibnik (Ithaca, N.Y.: Cornell University Press, 1987), 29–58. See also Tracy Bachrach Ehlers, "The Matrifocal Farm," in *Farmwork and Fieldwork*. Marty Strange tackles similar issues from a different perspective in *Family Farming* (Lincoln: University of Nebraska Press, 1988).

CHAPTER 4
Why Shrimpers Fail

1. See "Fisheries of the United States, 1991," National Marine Fisheries Service, NOAA, USDC (Washington, D.C.: GPO, 1992), xi.

2. The Alaskan fishery has been ranked number one for many years (see ibid., selected years).

3. Ibid., xii.

4. See Michael G. Haby and Richard E. Tillman, "The Texas Shrimp Industry: A Briefing Report," College Station, Texas Marine Advisory Service, July 1992; Texas Coastal and Marine Council, *Texas Bay Shrimp Industry* (Austin: Texas Coastal and Marine Council, 1983), 34–37; Texas Parks and Wildlife Department, *The Texas Shrimp Fishery: A Report to the Governor and the 72nd Legislature* (Austin: Texas Energies and Resources Council, 1991); Texas Energy and Natural Resources Council, "Coastal Natural Resources in Texas: Report to the Governor and the 68th Legislature, 1982," in *Policy Research Project Report* (Austin: LBJ School of Public Affairs, 1982). There are a number of problems with the accuracy of these figures; for example, all shrimp are not accounted for at the dock and therefore do not appear in any statistical profile; corresponding statistics for the Texas Gulf shrimpers are also problematic. See Harold F. Upton et al., "The Gulf of Mexico Shrimp Fishery: Profile of a Valuable National Resource," Center for Marine Conservation, Washington, D.C., 1992, 32–34.

5. See Texas Coastal and Marine Council, *Texas Bay Shrimp Industry,* 34.

6. See Haby and Tillman, "Texas Shrimp Industry," 4.

7. Ibid.

8. See Edward F. Klima et al., "Workshop on Definition of Shrimp Recruitment Overfishing," National Marine Fisheries Service, NOAA, USDC, Washington, D.C., 1990, and Klima, "Approaches to Research and Management of U.S. Fisheries for Penaied Shrimp in the Gulf of Mexico," in *Marine Invertebrate: Their Assessment and Management,* ed. John F. Caddy (New York: John Wiley and Sons, 1989), 87–113.

9. See Texas Coastal and Marine Council, *Texas Bay Shrimp Industry,* 36–39; Texas Parks and Wildlife Department, *Texas Shrimp Fishery: Report, 1991,* 2; and Haby and Tillman, "Texas Shrimp Industry: A Briefing Report."

10. Texas Coastal and Marine Council, *Texas Bay Shrimp Industry,* 38.

11. "Fisheries of the United States," selected years. See also John Vondruska, "Southeast Shrimp Fishery Market Conditions, 1991–1992," preliminary draft, 1992, National Marine Fisheries Service, St. Petersburg, Fla.; idem, *The Gulf Shrimp Industry* (St. Petersburg, Fla.: National Marine Fisheries Service, 1987); idem, "World Shrimp Situation 1990: Effects on Southeast Harvesters," *NOAA Technical Memorandum,* NMFS-SEFC-294, Galveston, Tex., 1991.

12. "Fisheries of the United States, 1986," 80.

13. Upton et al., "Gulf of Mexico Shrimp Fishery," 24.

14. Robert Lee Maril, *Texas Shrimpers* (College Station: Texas A & M University Press, 1983), 146–47.

15. See Texas Parks and Wildlife Department, *Texas Shrimp Fishery*, 6–7, and Texas Coastal and Marine Council, *Texas Bay Shrimp Industry*, 23. The number of boat licenses and corresponding fishermen must be regarded with a healthy degree of skepticism (see chapter 1, n.7, of this book).

16. The motivations and decisions of Texas lawmakers are more closely examined in the following chapters.

17. For a detailed explanation of these events and their consequences, see chapter 8.

18. J. A. Gulland, *Fish Stock Assessment*, as cited in Upton et al., "Gulf of Mexico Shrimp Fishery," 223.

19. See Klima, "Approaches to Research and Management"; Klima et al., "Workshop on Definition of Shrimp Recruitment Overfishing"; and Upton et al., "Gulf of Mexico Shrimp Fishery."

20. Joseph H. Kutkuhn, "Gulf of Mexico Commercial Shrimp Populations: Trends and Characteristics, 1956–1959," *Fishery Bulletin* 212, 62. U.S. Fish and Wildlife Service, Washington, D.C. (1962): 307–12.

21. William C. Gillespie, James C. Hite, and John S. Lytle, "An Econometric Analysis of the U.S. Shrimping Industry," *Economics of Marine Resources* 2, Department of Agricultural Economics and Rural Sociology, Clemson University (1969). See also Bruce A. Cox, "Statement before the United States International Trade Commission, 1976."

22. Cox, "Statement," 3.

23. Connor Bailey, "Social and Environmental Impacts of Shrimp Aquaculture in Indonesia," paper presented at the annual meeting of the Rural Sociological Society, Columbus, Ohio, 1991.

24. Walter R. Keithly, Jr., et al., "Effects of Shrimp Aquaculture on the U.S. Market: An Econometric Analysis," typescript, Center for Wetland Resources, Louisiana State University, Baton Rouge, 1992.

25. Ibid, 27–31.

26. See U.S. International Trade Commission, "Conditions of Competition Affecting the U.S. Gulf and South Atlantic Shrimp Industry" (Washington, D.C.: GPO, 1985).

27. For example, shrimp processors in Alabama were outraged when 600,000 pounds of imported shrimp were seized from them by the U.S. Customs Service. The processors argued that the improperly imported shrimp was in fact an American product because they had

deveined and packaged the shrimp. See Brad Clemenson, "When Is Foreign Shrimp a Product of the USA?" *Mobile Register*, June 10, 1988.

28. See chapter 6 for a more detailed discussion of these events.

29. See Keithly, Jr., et al., "Effects of Shrimp Aquaculture on the U.S. Market," 28–29.

30. Ibid.

31. This discussion relies heavily upon Connor Bailey and his work on shrimp production in Asia: see "Social and Environmental Impacts of Shrimp Aquaculture in Indonesia," Bailey and Mike Skladany, "Aquacultural Development in Tropical Asia," *Natural Resources Forum* 15, 1 (February 1991): 66–73, and Bailey, "The Social Consequences of Tropical Shrimp Mariculture Development," *Ocean and Shoreline Management* 11 (1988): 31–44.

32. Bailey and Skladany, "Aquaculture Development," 66–73.

33. Bailey, "Social Consequences," 31–49.

34. See Robert R. Stickney and James T. Davis, "Aquaculture in Texas," Texas A & M University Sea Grant Program, TAMU-SG-81-119, College Station, August 1981. See especially Addison L. Lawrence et al., "Shrimp Mariculture," Texas A & M University Sea Grant Program, TAMU-SG-82-503, College Station, 1981.

35. Pinkerton, "Pacific Shrimp in Laguna Madre," Coastal Conservation Association, Houston, *Tide*, March 1992, 3.

36. See Maril, *Texas Shrimpers*, 156–60.

37. E. Paul Durrenberger documents price fixing and its impact on shrimpers in Mississippi: see "Shrimpers, Processors, and Common Property in Mississippi," *Human Organization* 53, 1 (Spring 1994): 74–82. For an example of a recent price-fixing case in the fishing industry, see John H. Kennedy, "Seafood Firms, Officers Accused of Price Fixing," *Boston Globe*, January 29, 1994, 7:5, and Paul Hemp, "The Unraveling of a New England Fish Scam," *Boston Globe*, November 3, 1991, 1:2. See also "From Boat to Buyer, Gulf Shrimp Can Be a Game of Chance," *Seafood Leader* (Winter 1984): 38–41, and Perry W. Pawlyk and Kenneth J. Roberts, "Forecasting U.S. Shrimp Prices: A Comparison of Three Different Models," Center for Wetland Resources, Louisiana State University, Baton Rouge, 1985. This topic is discussed to a degree in Charles M. Adams and Fred J. Prochaska, "Principal Economic Factors Determining U.S. Shrimp Prices at Alternative Market Levels," Food and Resource Economics Department, University of Florida, 1984. Charles M. Adams et al. conclude that "no evidence was found to support the presence of monopsonistic pricing in the market" (see "Price Determination in the U.S. Shrimp Market," *Southern Journal of*

Agricultural Economics (December 1987): 103–11. I am grateful to Michael Haby for his helpful explanations on price structuring in the shrimping industry.

38. I would like to thank several informants for their insights into the Brownsville bid and price fixing. John Vondruska was especially helpful in providing price information and other data on imported shrimp.

39. The role of the multinational fish companies in foreign shrimp operations remains a rich topic for researchers, nor is much known about domestic fish processors. See Durrenberger, "Shrimpers, Processors, and Common Property in Mississippi."

40. See the discussion of candy in Maril, *Texas Shrimpers*, 163–64. Interviews with key informants in 1994 also suggest that kickbacks are still a part of the shrimping industry.

CHAPTER 5
The Role of the State

1. See T. R. Fehrenbach, *Lone Star*, (New York: Macmillan, 1975), 132–74.

2. See Robert Lee Maril, *Cannibals and Condos* (College Station: Texas A & M University Press, 1986). See also Martin Salinas, *Indians of the Rio Grande Delta* (Austin: University of Texas Press, 1990), and Thomas R. Hester, "Hunters and Gatherers of the Rio Grande Plain and the Lower Coast of Texas," Center for Archaeological Research, San Antonio, 1976.

3. See Lorraine Bruce Jeter, *Matagorda Early History* (Baltimore: Gateway Press, 1974), as cited in Keith Guthrie, *Texas Forgotten Ports*, (Austin: Eakin Press, 1988), 122.

4. See Guthrie, *Texas Forgotten Ports*, 124–25.

5. Ibid., 157–58.

6. Ibid., 69–73.

7. By 1920, with the addition of Kennedy, Kleberg, and Willacy counties, there were eighteen coastal counties in Texas.

8. For greater detail, see Robert Lee Maril, *Poorest of Americans* (Notre Dame, Ind.: University of Notre Dame Press, 1989), 19–34.

9. I have relied heavily on Seth Macinko, "Gospels of Efficiency, Public Policy and Rural Poverty: The Case of the U.S. Commercial Fishing Industry," paper presented at the annual meeting of the Rural Sociology Society, Columbus, Ohio, 1991.

10. Ibid., 4.

11. Ibid., 5.

12. Ibid., 3.

13. See Charles H. Stevenson, "Report on the Coast Fisheries of Texas," U.S. Commission of Fish and Fisheries, Report of the Commission, 1889–1891 (Washington, D.C.: GPO, 1893).

14. Ibid., 38.

15. Ibid.

16. Ibid. The hatchery is now used to raise Ridley sea turtles.

17. Ibid., 404.

18. See Stevenson, "Report on the Coast Fisheries of Texas." My review of the reports to the U.S. Commission of Fish and Fisheries, dating from 1871 to 1893, reveals no other documents that focus directly on Texas fisheries. Stevenson himself references two reports, "The Fisheries and Fishery Industries of the United States," 7 vols. (Washington, D.C.: GPO, 1881–1887), and "A Statistical Report on the Fisheries of the Gulf States," *Bulletin of the U.S. Fish Commission, 1891* (Washington, D.C.: GPO, 1892). The former, however, combines Texas data with aggregate data from other Gulf states and is thus of little use here; the latter's importance is superseded by Stevenson's much more comprehensive study.

19. See Stevenson, "Report on the Coast Fisheries of Texas," 38.

20. See Matagorda County Historical Commission, *Historic Matagorda County*, 2 vols. (Houston: O. Armstrong Company, 1986), and Stevenson, "Report on the Coast Fisheries of Texas," 393–96.

21. Stevenson, "Report on the Coast Fisheries of Texas," 375.

22. Matagorda County Historical Commission, *Historic Matagorda County*, 430.

23. Stevenson, "Report on the Coast Fisheries of Texas," 37.

24. Ibid., 404.

25. Ibid., 378.

26. Ibid., 375.

27. Ibid., 404.

28. Ibid., 405.

29. Ibid., 403.

30. Data are from Davor Jedlicka's demographic analysis of coastal counties in Texas (see "Population Changes in Texas Coastal Counties," typescript, Dept. of Sociology, University of Texas, Tyler). See also Calhoun County Historical Commission, *The Shifting Sands of Calhoun County, Texas* (Port Lavaca, Tex.: Calhoun Co. Historical Commission, 1981).

31. Stevenson, "Report on the Coast Fisheries of Texas," 403.

32. Ibid., 404.

33. Ibid., 395.

34. Ibid., 405.

35. Ibid., 402.

36. Ibid., 413.

37. Ibid., 405, 413, 415.

38. See Lewis Radcliffe, "Report Appendix X, Fishery Industries of the United States," *Report of the Division of Statistics and Methods of the Fishery for 1919*, Bureau of Fisheries doc. 892 (Washington, D.C.: 1920).

39. Ibid., 182–84.

40. Matagorda County Historical Commission, *Historic Matagorda County*, 439.

41. Ibid., 430.

42. Ibid., 420–31.

43. Stevenson, "Report on the Coast Fisheries of Texas," 375.

44. Radcliffe, "Report Appendix X," 180.

45. E. Paul Durrenberger documents this same process in Alabama in *It's All Politics* (Urbana: University of Illinois Press, 1992).

46. See Aransas Pass Chamber of Commerce, *Aransas Pass, Texas* (Aransas Pass, Tex.: Biography Press, 1976), 37–60.

47. Stevenson, "Report on the Coast Fisheries of Texas," 409.

48. See Fred F. Johnson and Milton J. Linder, "Shrimp Industry of the South Atlantic and Gulf States," USDC, Bureau of Fisheries report no. 21 (Washington, D.C.: GPO, 1934), 20–22.

49. Ibid., 30.

50. Ibid., 9.

51. Jedlicka, "Population Changes in Texas Coastal Counties."

52. Johnson and Linder, "Shrimp Industry of the South Atlantic and Gulf States," 61.

53. See Milton J. Linder and William W. Anderson, "Growth, Migration, Spawning, and Size Distribution of Shrimp, Penaeus Setiferus," *Fishery Bulletin* 56, 106 Fish and Wildlife Service, Washington, D.C. (1956): 555–64. The authors outline the reasons why coastal fisheries were restudied in the early 1930s.

54. Ibid., 15.

55. Johnson and Linder listed 428 fisherman as "on boats and shore." Seine shrimp fishing was accomplished, historically, by a minority of the shrimp fishermen. Estimating that minority at 10 percent of the work force, or 43 men, 385 men remain who worked in crews of two on board the smaller bay boats.

56. A similar method of estimation was again used, based upon the assumption that 10 percent of the work force were seiners.

57. Matagorda County Historical Commission, *Historic Matagorda County*, 431.

58. Ibid.

59. Johnson and Linder, "Shrimp Industry of the South Atlantic Gulf States," 23.

60. Ibid., 24. See also Robert Lee Maril, *Texas Shrimpers* (College Station: Texas A & M University Press, 1983), 170. Research on the historical and contemporary dimensions of this fish house work force is sorely needed.

61. Johnson and Linder, "Shrimp Industry of the South Atlantic and Gulf States," 25.

62. Ibid.

63. Ibid.

64. I would like to thank Emilio Zamora and Gary Libecap for their views of fishermen's unions in Texas (personal communications, April 10, 11, 1994).

65. See E. Paul Durrenberger, "Psychology, Unions, and the Law: Folk Models and the History of Shrimpers' Unions in Mississippi," paper presented at the annual meeting of the Association for Applied Anthropology, Charleston, S.D., 1991.

66. Johnson and Linder, "Shrimp Industry of the South Atlantic and Gulf States," 15. To calculate this estimate of men, I first subtracted 10 percent from the total number of fishermen, my estimate of the number of seine fishermen on shore who fished and who were included in the total.

67. I took the total number of men on motorboats, 530, then divided by average crew, two per boat, to arrive at the number of boats. I did not figure in the number of seine fishermen at 10 percent because by 1927 seine fishing was rare.

68. Johnson and Linder, "Shrimp Industry of the South Atlantic and Gulf States," 15.

69. I examined issues of *American Federationist* for an overview of annual activities of the American Federation of Labor and Congress of Industrial Organizations (AFL-CIO) from 1921 to 1939. For a general critique of the AFL-CIO, see Philip Taft, *The A. F. of L. in the time of Gompers* (New York: Harper, 1957), and idem, *The A. F. of L. from the Death of Gompers to the Merger* (New York: Harper, 1959).

70. See Ronald N. Johnson and Gary Libecap, "Regulatory Arrange-

ments and Fishing Rights in the Texas Bay Shrimp Fishery," typescript, Department of Economics, Texas A & M University.

71. Ibid., 23–29.

72. Ibid., 26.

73. Ibid., 27.

74. Ibid.

75. See Durrenberger, "Psychology, Unions, and the Law," 3.

76. Ibid., 3, 4, 6.

77. Johnson and Libecap, "Regulatory Arrangements and Fishing Rights," 24.

78. Columbia River Packers v. Hinton, 315 U.S. 520 (1942), and Manaka v. Monterey Sardine Industries 41 JF. Supp. 531 (1941), ibid.

79. Durrenberger, "Psychology, Unions, and the Law," 4.

80. Johnson and Libecap, "Regulatory Arrangements and Fishing Rights," 18–22, as well as Durrenberger, "Psychology, Unions, and the Law," 3, discuss this important case.

81. See Johnson and Libecap, "Regulatory Arrangements and Fishing Rights," 25, and Durrenberger, "Psychology, Unions, and the Law," 4.

82. Johnson and Libecap, "Regulatory Arrangements and Fishing Rights," 25.

83. I found no archival evidence for unions in Texas, and there was no mention of union activities among Texas fishermen in *American Federationist* from 1921 to 1939. Neither did I find in any of the Texas coastal county libraries any documentary evidence suggesting the existence of bay shrimper unions. And Libecap, who interviewed bay shrimpers specifically about unions, found no evidence suggesting there were unions of bay shrimpers in Texas. It may well be that in a few of the fishing communities bay shrimpers organized, but the lack of any supporting evidence suggests it is doubtful.

84. Johnson and Linder, "Shrimp Industry of the South Atlantic and Gulf States," 61.

85. Ibid., 61, 62.

86. Johnson and Libecap, "Regulatory Arrangements and Fishing Rights," 40.

87. Ibid., 40, 39.

88. See Maril, *Texas Shrimpers*, 75–97.

89. Data are based on interviews with a Brownsville Gulf shrimper who had fished for several years during the 1950s in Bluefields, Nicaragua; he escaped with his family in his trawler when the Sandinistas came to power.

90. Data are based upon 1979 interviews with Texas Gulf shrimpers in Port Isabel and Brownsville and from informal discussions with their wives.

91. I am grateful to Virginia Voltaggio Wood for providing information about the history of her family.

92. Data are taken directly from Jedlicka, "Population Changes in Texas Coastal Counties," Appendix, Table 1.

93. Maril, *Texas Shrimpers,* 76–77.

94. Ibid., 75–97.

95. Ibid., 81.

96. Aransas Pass Chamber of Commerce, *Aransas Pass, Texas,* 141.

97. Maril, *Texas Shrimpers,* 82.

98. Given historical hindsight, one would question the wisdom of selecting Aransas Pass and other coastal communities for the resettlement of Vietnamese refugees who intended to shrimp only twenty-five years or so after Cajun shrimpers in Aransas Pass were run out of town (see chapter 8 for further discussion).

99. Johnson and Libecap, "Regulatory Arrangements and Fishing Rights," 37.

100. Ibid., 36–37.

101. Unions were outlawed in 1956 in "Gulf Coast Shrimpers and Oysterman's Association v. U.S., no. 156680, 5th Cir., U.S. Court of Appeals, 1956; 51–53.

102. Johnson and Libecap, "Regulatory Arrangements and Fishing Rights," 41–42.

103. Earl Thompson's novel *Caldo Largo* (New York: New American Library, 1976) is in part a fictionalized account of this cat-and-mouse game between Texas shrimpers and Mexican gunboats; the same game is played in the 1990s (see, for instance, James Pinkerton, "Shrimper: Mexicans Shot at Ship—U.S. Tried to Pinpoint Location of Incident," *Houston Chronicle,* March 31, 1992, 9).

104. For a more detailed description, see Maril, *Texas Shrimpers,* 187–89.

105. There were probably smaller strikes in other Texas ports. A complete study of Gulf shrimpers' unions along the coast of the Gulf of Mexico awaits scholarly research.

106. Maril, *Texas Shrimpers,* 165–67.

CHAPTER 6
New Alliances

1. See Estellie M. Smith, "The Public Face of the New England Regional Fishery Council: Year 1," typescript, 1978.

2. See Richard Travis Christian, "Decision-Making in Coastal Fish-

eries Conflict: The Case of Red Drum and Spotted Seatrout Legislation in Texas," Master's thesis, Texas A & M University, 1986, 45.

3. Charles H. Stevenson, "Report on the Coast Fisheries of Texas," U.S. Commission of Fish and Fisheries, Report of the Commission, 1889–1891 (Washington, D.C.: GPO, 1893), 379.

4. For many of the details of this discussion I rely on the work of Christian, "Decision-Making in Coastal Fisheries Conflict." The conceptual analysis, the conclusions drawn from it, and the placement of the analysis within a larger sociopolitical framework are mine.

5. See Robert B. Ditton, Anthony J. Fedler, and Richard T. Christian, "The Evolution of Recreational Fisheries Management in Texas," typescript, May 1991, Texas A & M University.

6. The Sportsmen Conservationists of Texas (SCOT) has 30,000 members and is larger than the Gulf Coast Conservation Association (GCCA); SCOT is affiliated with the National Wildlife Federation.

7. Christian, "Decision-Making in Coastal Fisheries Conflict," 60–61.

8. Peter Elkind, "The Buck Stops Here," *Texas Monthly* (January 1990): 100–105.

9. See Ditton et al., "Evolution of Recreational Fisheries Management in Texas," 5–6.

10. Christian, "Decision-Making in Coastal Fisheries Conflict," 118–20.

11. Ibid., 51–56, 122–31.

12. Ibid., 88–89.

13. Ibid., 129–30.

14. Ibid., 85–86.

15. Ibid., 110.

16. Dick J. Reavis, "The New Rustlers," *Texas Monthly* 2, 4 (1983): 81–84.

17. Ibid.

18. Christian, "Decision-Making in Coastal Fisheries Conflict," 124.

19. Robert Lee Maril, *Texas Shrimpers* (College Station: Texas A & M University Press, 1983), 180–81.

20. The TPWD decided that East Matagorda Bay would remain closed for 120 days, becoming a fishery sanctuary for drum and trout. It then announced that it would keep the bay closed indefinitely. As was the case with the data presented for the ten bills on drum and trout, the department mounted a presentation of biological data to support its decision to keep the fishery closed to recreational fishermen. A crucial part of their argument was that the recreational fishery

in the bay was economically insignificant. The agency determined that the economic impact of the closed fishery therefore would be negligible. A series of public hearings were held before the TPWD's decision became law, and the response was loud and clear, with more than 600 people turning out in Bay City to oppose the department's decision. Recreational fishermen were outraged, and local businessmen, who depended on their dollars, were also angered. Responding to the political pressure, the TPWD reversed its decision, at the same time denying that it had ever seriously considered a permanent closure of East Matagorda Bay. In "Sport Fishery Management in East Matagorda Bay (Texas): An Analysis of Decision Making" (Master's thesis, Texas A & M University, 1991), Mary Christine Ritter notes that the TPWD, sensing the public pressure it was about to face, changed its mind about the closing before the public hearings ever took place.

21. Material is based on interviews with boat owners in Brownsville and Port Isabel in 1982 and 1983 and on firsthand observation of the series of events described. Data from my wage studies of Brownsville and Port Isabel shrimpers prior and subsequent to the enforcement of the Lacey Act lend additional support to my conclusions. Additional interviews were conducted with the U.S. Coast Guard, National Marine Fisheries Service, and other people involved. I also became an unwilling participant in events surrounding the enforcement of the Lacey Act, first as a researcher, then as a legal expert witness, and finally as a recipient of political pressure. These roles and experiences allowed me further observation of and insight into the Lacey Act, its enforcement, and its politics. I am indebted to several members of the Brownsville legal community for their insightful interpretations of law. See "They Followed Us with Jets and My Own Country Arrested Me: The Arbitrary Enforcement of the Lacey Act," a paper I presented at the annual meeting of the Rural Sociology Society, College Station, Texas, 1984, and "Unsolicited Report," findings I submitted to the National Marine Fisheries Service in 1983.

22. Along the U.S.-Mexican border a variety of laws must be overcome daily in order to conduct normal business between the two nations. It is not at all unusual for businesses to overlook laws that are considered to be extraneous and for law enforcement agencies to do likewise. See, for instance, *Borderlands Sourcebook*, ed. Ellwyn R. Stoddard et al. (Norman: University of Oklahoma Press, 1983), 214–16.

23. See Lacey Act, 16 U.S. Code 3401-3408, Public Law 97-79. See also House of Representatives Lacey Act amendments of 1981, report no. 97-276, 97th Cong., 1st Sess.; Lacey Act implementation, memo-

randum to William G. Gordon from Stephen J. Powell, February 17, 1982; and formal legal opinion, Office of the General Counsel, NOAA no. 69, Application of the Black Bass Act, 16 U.S.C. 852-856, to purchasers or receivers of Puget Sound salmon, by A. R. Watson, August 30, 1978.

24. JR 5604, the amended Lacey Act, was drafted by the Carter administration in 1979, and hearings were held by the Subcommittee on Fisheries and Wildlife Conservation and the Environment. The bill received support from the Merchant Marine and Fisheries Committee and passed the House in July 1980 but was not acted upon by the Senate. An identical bill, JR 1638, was sponsored by John Breaux, then congressman from Louisiana. Hearings were again held, and this time the bill was passed by the House and the Senate and signed into law.

25. The timing of the Lacey Act and why it was amended to include shrimp are beyond the range of my topic, but the issues offer a fertile subject for future study. Research would necessarily focus on the long-time conflict between Louisiana and Texas Gulf shrimpers and their ability to mobilize elected officials.

26. My report on the Lacey Act resulted in a classic example of killing the messenger. I submitted the report to the National Marine Fisheries Service, and it brought disfavor upon me and upon the funding agency that supported my research; I was told in so many words to mind my own business. My professional competence as a sociologist was called into question, and my unsolicited report to the NMFS was discredited. Subsequent funded research by other social scientists documented the veracity of my report, however (see Ben M. Crouch and Mark K. Miller, "Lacey Act Enforcement in the Texas Gulf: A Sociological Analysis," in *Proceedings of the Conference on Gulf and South Atlantic Fisheries: Law and Policy*, ed. M. C. Jarman and D. K. Conners, Mississippi-Alabama Sea Grant Program, 87-130, New Orleans, 1987, 121–35). I especially would like to thank an anonymous informant for his information on this topic.

27. See chapter 2.

28. For a discussion of the impact of investor boats on the shrimping industry, see Maril, *Texas Shrimpers*, 146–48. The government financed commercial fishing boats in 1936 under the Fisheries Obligation Guarantee Program, part of the Merchant Marine Act; in the 1970s some shrimpers took advantage of the "7/11" loan, a 7 percent interest loan for eleven years. Both programs were operated through the NMFS.

29. See Peggy Fikac, "Shrimp, Oyster Rules Will Be Under Wildlife Department," *Brownsville Herald*, April 30, 1985, 9 (emphasis added).

30. Ibid.

31. Ibid.

32. See *Texas Shores* 23, 3 (1990): 21.

33. For a discussion of TEDs and Alabama shrimpers, see E. Paul Durrenberger, "Shrimpers and Turtles on the Gulf Coast: The Formation of Fisheries Policy in the United States," *Maritime Anthropological Studies* 1, (1988): 196–214. For an overview of TEDs and Mexican shrimpers, see Jack Rudloe and Anne Rudloe, "Shrimpers and Lawmakers Collide over a Move to Save the Sea Turtles," *Smithsonian*, December 1989, 45–54.

34. Michael K. Orbach notes in his ethnography of California tuna fishermen that they had no desire to net dolphins; but unlike the situation with shrimpers who caught sea turtles, the attempt to free dolphins from tuna nets was dangerous (see Orbach, *Hunters, Seamen, and Entrepreneurs: The Tuna Seinermen of San Diego* (Berkeley: University of California Press, 1977).

35. Stevenson, "Report on the Fisheries of Texas," 411–12.

36. Ibid., 412.

37. Texas shrimpers interviewed in the sample reported that Mexican shrimpers actively fished for sea turtles because of the high price they brought in Mexican ports as food and as an aphrodisiac.

38. Stevenson, "Report on the Fisheries of Texas," 412.

39. Rudloe and Rudloe, "Shrimpers and Lawmakers Collide," 47.

40. A full-fledged study of the development of TEDs and its impact upon the environmental movement and the shrimping industry, a topic for future research, would reveal much about the relationships among local regional and national interests, the various levels of government, and private industry.

41. See Durrenberger, "Shrimpers and Turtles on the Gulf Coast," 201–2.

42. For a description of the political fray through the eyes of environmentalists and of fishermen, see Marydee Donnelly, "New TED Regulations Due Out after Months of Delay," *Marine Conservation News* (Spring 1992): 15; idem, "National Academy of Sciences Recommends More TED Regulations," *Marine Conservation News* (Fall 1990): 3; Brad Clemenson, "Judge Backs TEDs Suspension," *Mobile Press Register*, July 29, 1989, 1A; Edith Gray and Richard Jensen, "Shrimping Industry vs. Feds," *Baldwin Today*, May 17, 1989, 1c; "Shrimpers Win Reprieve From Turtle Device Laws," *Texas Shoreline* 1, 3 (1988): 1; "Shrimpers Ordered to Use Trap Door in Their Nets," *Tulsa World*, Sep-

tember 6, 1989, 14B; and "TEDs Agreement Prompts Senator to Drop Protests," *Brazosport Facts*, July 15, 1988, 9.

43. In general, the media underplayed the extent of the protest and the intensity of the fishermen's reaction to the enforcement of TEDs (see, for instance, "Coast Guard: Shrimpers Ram Supply Boat in TEDs Protest," *Mobile Press Register*, July 23, 1989, 4b).

44. Clemenson, "Judge Backs TEDs Suspension," and "Shrimpers Ordered to Use Trap Door in Their Nets."

45. Donnelly, "New TED Regulations."

46. See, for example, Debby Crouse, "TEDs! At Long Last!" *Marine Conservation News* 5, 1 (Spring 1993): 1.

47. Donnelly, "New TED Regulations Due Out after Months of Delay," 15.

48. *Texas Shores* 23, 3 (Fall/Winter 1990): 21.

49. *American Heritage Dictionary,* High School ed., s.v. "turtle."

50. Anna Mearns, "The Adventures of Ranger Rick," *Ranger Rick* (June 1992): 18.

51. National Academy of Sciences, "The Decline of Sea Turtles: Causes and Preventions," May 1990.

52. Crouse, "TEDs! At Long Last!"

53. See, for example, Concerned Shrimpers of America Inc., v. Florida Marine Fisheries Commission and Center for Marine Conservation; Florida Audubon Society; and Greenpeace, U.S.A. Case no. 89-4220R, State of Florida Division of Administrative Hearings, Tallahassee, Sept. 8, 1989. In this case, as in the others I observed, the bay shrimpers were always outmaneuvered by superior legal counsel. Advocates for TEDs included the best lawyers money could buy and government attorneys with special expertise; legal counsel for shrimpers had much less expertise and was often not as well prepared.

54. See, for instance, Pamela Casteel, "By-Catch the Nineties Issue," *Texas Shores* (Fall/Winter 1991): 2. BEDs have also been called BRDs (by-catch reduction devices). See Sonja Fordham, "CMC Lone Environmental Voice at International Shrimp By-catch Conference," *Marine Conservation News* 4, 3 (Autumn 1992): 18.

55. See, for example, Sonja Fordham, "CMC's First Shrimp By-Catch Workshop a Huge Success," *Marine Conservation News* 4, 3 (Spring 1992): 18.

56. Harry Upton, "CMC to Convene Workshop on Incidental Catch in Shrimp Fishery," *Marine Conservation News* (Winter 1991): 9.

57. See, for example, Pamela Casteel, "Snapper Fishing for Fun: Is It Worth It?" *Texas Shores* (Fall/Winter 1991): 9.

58. Mearns, "Adventures of Ranger Rick," 20.

59. Transcript, *Meeting of the Shrimp Advisory Committee*, Austin, Texas Park and Wildlife Department, December 2, 1991, 167.

CHAPTER 7
Women Who Shrimp the Bays

1. This chapter is based in part on Andrea Fisher Maril, "Women and Shrimping: The Texas Way," paper presented at the annual meeting of the Society for Applied Anthropology, San Antonio, Texas, March 1993.

2. Estellie Smith points out that the lengthy separations created by offshore fishing often contribute to the development of more independence for women who are married to shrimpers (see Smith, ed., *Those Who Live from the Sea* [St. Paul, Minn.: West Publishing Company, 1977]).

3. Dona Davis uses the term to describe fishermen's wives who take responsibility for many of the land-based chores usually accomplished by their husbands; see Davis, " 'Shore Skippers' and 'Grass Widows': Active and Passive Roles for Women in a Newfoundland Fishery," in *To Work and to Weep: Women in Fishing Economies*, ed. Jane Nadel and Dona David (St. John's: Institute of Social and Economic Research, Memorial University of Newfoundland, 1988). With the exception of the Pacific Northwest, women boat captains are seldom found in American commercial fisheries (see Davis and Nadel's introductory chapter for an excellent review of the literature on women who fish).

4. Using a snowball-sample technique, I interviewed nineteen women in Freeport and Sargent, Texas. In-depth interviews were conducted in fish houses, bait shops, on the boat and docks, and in the homes of respondents; the interviews usually took about an hour to complete although some took much longer. In addition to being questioned about demographic data, women were asked to discuss how and why they began to shrimp, the nature of their work on the bay, and their job satisfaction. The female fishermen were also asked about the relationship between shrimping and family as well as their involvement in the community and their political concerns. Finally, they were asked to reflect upon the problems and issues they may have faced as women working in a male occupation and work setting. Norman K. Denzin discusses this sampling technique as a means from moving

from a random to a nonrandom sample (see *The Research Act: A Theoretical Introduction to Sociological Methods*, 3d ed. [Englewood Cliffs, N.J.: Prentice-Hall, 1989]). Here I follow the approach of Catherine Kohler Reissman in her study of divorcing individuals. In her sampling strategy, the first set of informants was located through probate court records of those people filing for divorce in two counties in the northeast. At the end of each interview, informants were asked for the names of others who fit the interview criteria. No more than one person named by each interviewee was interviewed for the final sample. The sampling frame was constructed in a similar fashion: ten women were interviewed in the original sample. At the end of the interview, each woman was asked to supply names of other women who shrimped. Nine of these women, no more than one from each original respondent, were subsequently interviewed for a total of nineteen interviews. See Catherine Kohler Reissman, *Divorce Talk: Women and Men Make Sense of Personal Relationships* (New Brunswick, N.J.: Rutgers University Press, 1990). Because of the way in which the questions in the survey were administered, it was impossible to determine the proportion of men whose wives or girlfriends had worked as fishermen, a problem future researchers will want to correct. Although the findings from this subsample of women who shrimp are admittedly tentative, given the recognized limitations of the methodology, they do raise a number of questions centered on gendered behavior and expectations that researchers should consider.

5. The emphasis on active rather than on shore-based roles for women in this subsample was not intended to downplay the more traditional, and more widespread, role of shrimpers' wives who remain on shore. Yet as Nadel and Davis note, "Female contributions are often overlooked or relegated to the more passive, landbound roles of reproduction and rearing of future generations of fishermen" (see Nadel and Davis, *To Work and to Weep*, 3).

6. Davis discusses a similar pattern in the New England fishery (see Davis, " 'Shore Skippers' ").

7. Germaine Greer examines this issue in *The Change: Women, Aging, and Menopause* (New York: Alfred A. Knopf, 1991). Sally Cole and Lila Abu-Lughod also note that women in other cultures gain freedom from gender-role constraints with the onset of menopause or after childbearing duties have been completed. See Cole, *Women of the Praia* (Princeton, N.J.: Princeton University Press, 1991), and Abu-Lughod, *Veiled Sentiments: Honor and Poetry in a Bedouin Society* (Berkeley: University of California Press, 1986).

8. Fran Danowski notes, for example, that "experience is the teacher that can never be replaced by classroom learning when it comes to things like reading weather signs, getting the feel of sea bottom conditions, and judging safe risks that could make a difference between subsistence and profit" (see "Fishermen's Wives: Coping with an Extraordinary Occupation," *University of Rhode Island Marine Bulletin 37*, NOAA/Sea Grant Program, Narragansett, R.I., 7).

9. Various observers have pointed out the relationship between the rise of a cult of domesticity and the separation of paid male laborers and unpaid female household workers. Women justified their existence based upon the value of their separate sphere centered on home, hearth, and children; they were different from men and valued this difference. Wives of affluent husbands stayed home and promoted their separate spheres; women who had to work were more likely to downplay their contribution to the household economy to avoid being seen as unwomanly. See Jesse Bernard, *The Female World* (New York: Free Press, 1981); Helene Z. Lopata, "The Interweave of Public and Private: Women's Challenge to American Society," *Journal of Marriage and the Family* 55 (1993): 176–90; and Carolyn E. Sachs, *The Invisible Farmers: Women in Agricultural Production* (Totowa, N.J.: Rowman and Allanheld, 1983). Note that Sachs argues that farm wives, like female shrimpers, have been invisible.

10. Males also use seasickness as a face-saving device to excuse themselves from the rigors of commercial fishing (see Robert Lee Maril, *Texas Shrimpers* [College Station: Texas A & M University Press, 1983], 4–6).

11. Maureen Perry-Jenkins et al. found similar patterns in their studies of women based upon whether they regarded themselves as primary, secondary, or ambivalent providers. Women who saw themselves as primary providers had fewer interruptions in their work histories. See Perry-Jenkins, Brenda Seery, and Ann C. Crouter, "Linkages between Women's Provider-Role Attitudes, Psychological Well-Being, and Family Relationships," *Psychology of Women Quarterly* 16 (1992): 311–29.

12. In examining men's participation in housework across occupations, Nancy C. Gunther and B. G. Gunther found that couples with more androgynous ideas about gender roles were more likely to share housework than couples with more traditional beliefs about the roles of men and women (see "Domestic Division of Labor among Working Couples: Does Androgyny Make a Difference?" in *Psychology of Women Quarterly* 14 [1990]: 355–70).

13. Lillian Breslow Rubin discusses much the same pattern among lower-class women; see *Worlds of Pain: Life in the Working-Class Family* (New York: Basic Books, 1976).

14. See Sherry B. Ortner and Harriet Whitehead, "Introduction: Accounting for Sexual Meanings," in *Sexual Meanings: The Cultural Construction of Gender and Sexuality*, ed. Ortner and Whitehead (Cambridge: Cambridge University Press, 1981).

15. See ibid. Dorinne E. Kondo suggests that the cultural construction of gender is context-specific, an ongoing negotiation among participants that is grounded in historical and cultural variables (see *Crafting Selves: Power, Gender, and Discourse of Identity in a Japanese Workplace* [Chicago: University of Chicago Press, 1990]). See also Susan Faludi, *Backlash: The Undeclared War against American Women* (New York: Crown Publishers, 1991).

16. Carolyn E. Sachs found that it was acceptable for a woman to help out her husband occasionally, but a woman whose husband needed her help too frequently lost status in the eyes of her peers (see Sachs, *Invisible Farmers*), 51–52.

17. Ibid. Sachs notes that in the 1960s women on farms began identifying themselves as farmers, not as helpers or farm wives. Her research suggests that this shift has little to do with a change in actual duties but instead reflects changes in how both husbands and wives define women's identity. Rachel Ann Rosenfeld finds this same dynamic; as women enter the work force in other occupations, it has become more acceptable for farm women to identify themselves as farmers (see Rosenfeld, *Farm Women: Work, Farm and Family in the United States* [Chapel Hill: University of North Carolina Press, 1985]. See also Wava G. Haney and Jane B. Knowles, *Women and Farming: Changing Roles, Changing Structures* (Boulder, Colo.: Westview Press, 1988).

18. Definitions of work have plagued attempts to understand the economic contributions of women. See Barbara Reskin and Irene Padavic, *Women and Men at Work* (Thousand Oaks, Md.: Pine Forge Press, 1994); Teresa Amott and Julie Matthaei, *Race, Gender, and Work: A Multicultural Economic History of Women in the United States* (Boston: South End Press, 1993); and Lourdes Beneria, "Accounting for Women's Work: The Progress of Two Decades," *World Development*, 20, 11 (1992): 1547–60. A special thanks to Elizabeth Howard for her insights into these issues.

19. Sandra Bem asserts that our culture views worth through an "androcentric" lens, deeming men's contributions as intrinsically more valuable than women's; paid work therefore is by definition more valu-

able than unpaid work (see *The Lenses of Gender: Transforming the Debate on Sexual Inequality* [New Haven, Conn.: Yale University Press, 1993]).

20. See Haney and Knowles, *Women and Farming*, and Rosenfeld, *Farm Women*.

21. See, for example, Glenda Riley, *The Female Frontier* (Lawrence: University Press of Kansas, 1988); Susan Armitage and Elizabeth Jameson, eds., *The Women's West* (Norman: University of Oklahoma Press, 1987); Norton Juster, *So Sweet to Labor: Rural Women in America, 1865–1895* (New York: Viking Press, 1979); and Glenda Riley, *A Place to Grow: Women in the American West* (Arlington Heights, Ill.: Harland Davidson, 1992).

22. Those few women who have identified themselves with the occupation of fishing have always preferred the term "fisherman" over "fisher," "fisherfolk," or "fisherwoman," a choice that may change as other linguistic conventions begin to reflect a growing awareness of how language shapes perception of gender.

23. Rosabeth Moss Kanter, *Men and Women in the Corporation* (New York: Basic Books, 1977).

CHAPTER 8
Vietnamese Bay Shrimpers

1. Although a majority of the refugees who moved to Texas fishing communities were Vietnamese, approximately 5 percent were Laotians or Cambodians. Texas fishermen, along with other residents of the communities, did not distinguish between the different nationality groups, referring to all refugee newcomers as "Vietnamese." See Louise Flippin, "The Indochinese Refugee Situation: America's Newest Wave of Immigrants," Center for Asian Studies, Austin, Texas, 1979.

2. A minority of the refugees identified themselves as ethnic Chinese, not Vietnamese. See William E. Spruce, "Indochinese Refugees on the Central Coast of Texas," typescript, U.S. Department of State, Foreign Service Institute, 1985. See also John D. Donohue, *Cam An: A Fishing Village in Central Vietnam* (East Lansing: Michigan State University Press, 1961); William Liu, Mary Anne Lamanna, and Alice Murata, *Transition to Nowhere: Vietnamese Refugees in America* (Nashville, Tenn.: Charter House, 1979); and Gail Paradise Kelly, *From Vietnam to America* (Boulder, Colo.: Westview Press, 1977). For another overview, see Darrell Montero, *Vietnamese Americans: Patterns of Resettlement and Socioeco-*

nomic Adaptation in the United States (Boulder, Colo.: Westview Press, 1979).

3. See Paul Starr, "Troubled Waters: Vietnamese Fisherfolk on America's Gulf Coast," *International Migration Review* 13 (1979): 25–45.

4. The analysis of Vietnamese refugees in Port Isabel is based upon data collected from a variety of sources. In 1979 I interviewed thirty Gulf shrimpers in Port Isabel, many of whom remembered the events in which the Vietnamese refugees participated. Later I discussed the problems of resettlement in Port Isabel with staff of one of their major sponsors, the Catholic Diocese of Brownsville. I also interviewed fish house operators in Port Isabel and Port Brownsville who were familiar with the events. In 1983 and 1984 I returned to study the community of Port Isabel (partial findings published in Robert Lee Maril, *Cannibals and Condos* [College Station: Texas A & M University Press, 1986). From 1985 to 1987 I was again in constant contact with the community through involvement in another study (see Maril, *Poorest of Americans* [Notre Dame, Ind.: University of Notre Dame Press, 1989], 77–78, 124–25).

5. See Maril, *Poorest of Americans*.

6. See Robert Lee Maril, *Texas Shrimpers* (College Station: Texas A & M University Press, 1983), 75–97.

7. My knowledge of the community of Seadrift began in 1978 when I interviewed shrimpers there and in nearby communities (see Maril, *Texas Shrimpers*, 183–85). I returned there in 1984 to interview leaders and fishermen (research summarized in Robert Lee Maril, "Five Years after a Death: Conflict and Assimilation in a Texas Coastal Community," American Fishery Society, Sun Valley, Idaho, 1985). I again interviewed bay shrimpers in Seadrift in 1988 and 1989, asking a series of questions about their attitudes toward the Vietnamese.

8. See the historical overview of Seadrift in chapter 5.

9. For a more detailed description, see Maril, "Five Years after a Death."

10. The facts about Aplin's death are based on John Bloom, "A Delicate Balance," *Texas Monthly* (October 1979): 124, 256–58; the transcript from the television program "20/20," which first aired September 20, 1979; and a number of published accounts, including Molly Ivins, "Killing Sharpens Texas Feud on Vietnamese Fishing," *New York Times*, August 9, 1979, A16, and Neil Matwell, "Texans Vexed at Vietnamese Who Work Harder Than They Do," *Asian Wall Street Journal*, August 11, 1979, 1. See also Trans-Century Corporation, report to Minority Business Development Agency, "New Economic and Social Opportunities

for Americans and Indochinese on the Texas Gulf Coast" (typescript), USDC, Washington, D.C., October 15, 1979. Three Seadrift residents were interviewed about the incident.

11. See Bloom, "A Delicate Balance," 69–70.

12. See Trans-Century Corporation report, "New Economic and Social Opportunities," for a detailed discussion.

13. See Clifford Cain, "Committee Backed," *Victoria Advocate*, September 12, 1979, 1A, 12A.

14. Quoted from transcript of "20/20," September 20, 1979, 10.

15. Ibid., 1.

16. See Ronald N. Johnson and Gary D. Libecap, "Regulatory Arrangements and Fishing Rights in the Texas Bay Shrimp Fishery," typescript, Dept. of Economics, Texas A & M University, 22–23, 49.

17. David McLemore, "A Cold War Heats Up along the Texas Coast," *San Antonio Express*, February 8, 1981, 1A, 6A. See also William Stevens, "Viet Spice in Houston," *New York Times* News Service, 1981.

18. The analysis of Palacios is based on six weeks of fieldwork there in 1988 and on two additional weeks in 1989 during which time I interviewed members of both the Vietnamese and the non-Vietnamese community. I owe a large debt of gratitude to Father Joseph Pham duc Trinh for his help and patience.

19. See Palacios Chamber of Commerce, "Palacios Chamber of Commerce Community Data Profile," 1986, and "Palacios Economy Is Reviewed by Diocese," *Palacios Beacon*, December 2, 1987, 2.

20. "Palacios Economy Is Reviewed by Diocese," 2.

21. I attended six different masses celebrated by the Vietnamese priest, each one conducted entirely in Vietnamese. Women were seated on the left side of the church, men on the right, as was the tradition in Vietnam, with younger family members in the front pews under the direct supervision of an adult. No written materials had been available to the congregation in Vietnam, no Bibles, books of religious readings, or hymnals; the oral responses, alternating between men and women, were an integral part of the service and had been memorized. The Vietnamese priest emphasized that some parts of the mass had been changed from the original inadvertently over time; he believed that his flock would resent any attempts to change their responses to a text more faithful to the original.

22. John Howarth, "Committee Studying Zoning Pros and Cons," *Palacios Beacon*, February 23, 1983, 1.

23. "Concern Voiced over Fire Safety in Boat Building," *Palacios Beacon*, March 21, 1984, 2.

24. See Howarth, "Committee Studying Zoning Pros and Cons," 1.

25. "Self-Regulation Goal Sought by New Bay Shrimp Organization," *Palacios Beacon*, February 22, 1984, 1.

26. "Bethany Park Subdivision Groundbreaking," *Palacios Beacon*, June 27, 1984, 1.

27. American students' acceptance of the Vietnamese in their schools requires much further study. Participation in school activities and above-average grades in themselves do not demonstrate that students do not suffer discrimination.

28. I interviewed fish house owners, marine agents, dockside workers, and other residents in Palacios about their attitudes toward the local Vietnamese fishermen. I also studied Vietnamese bay shrimpers in Bayou La Batre, Alabama, during fall 1989; several of the shrimpers there had lived in Texas fishing communities, including Seadrift and Palacios. I interviewed eighteen Vietnamese bay shrimpers in Bayou La Batre, using a structured, in-depth interview that lasted from forty minutes to an hour and a half. I am indebted to Mr. Pham Dinh Lan of the Catholic Refugee Settlement Program in Bayou La Batre for his valuable help as translator and for sharing his knowledge about Vietnamese fishermen in his community. A detailed ethnographic study of Texas Vietnamese fishermen is a subject that researchers should seriously consider.

29. See, for example, Lloyd Grove, "Vietnamese Not Rushing for Residency," *Dallas Morning News*, June 19, 1978, 1C, and Lilla Rose, "Diocese Cautions Vietnamese Priest," *Beaumont Express*, May 11, 1977, 7A.

SELECTED BIBLIOGRAPHY

❖   ❖   ❖

Abrahams, Roger, D. 1974. *Deep the Water, Shallow the Shore*. Austin: University of Texas Press.

Abu-Lughod, Lila. 1986. *Veiled Sentiments: Honor and Poetry in a Bedouin Society*. Berkeley: University of California Press.

Acheson, James. 1981. Anthropology of Fishing. *Annual Review of Anthropology* 10: 275–316.

Allen, Charles W. 1981. Gulf Shrimping and Technology. Department of History, Texas A & M University.

Amott, Teresa, and Julie Matthaei. 1993. *Race, Gender, and Work: A Multicultural Economic History of Women in the United States*. Boston: South End Press.

Aransas Pass Chamber of Commerce, 1976. *Aransas Pass, Texas*. Aransas, Texas: Biography Press.

Armitage, Susan, and Elizabeth Jameson, eds. 1987. *The Women's West*. Norman: University of Oklahoma Press.

Avioli, Paula Smith, and Eileen Kaplan. 1992. A Panel Study of Married Women's Work Patterns. *Sex Roles* 26 (5–6): 227–41.

Bailey, Connor. 1988. The Social Consequences of Tropical Shrimp Mariculture Development. *Ocean and Shoreline Management* 11: 31–44.

———. 1991. Social and Environmental Impacts of Shrimp Aquaculture in Indonesia. Paper presented at the annual meeting of the Rural Sociological Society, Columbus, Ohio.

Bailey, Connor, and Mike Skladany. 1991. Aquacultural Development in Tropical Asia. *Natural Resources Forum* 15(1): 66–73.

Belshaw, Cyril S. 1965. *Traditional Exchange and Modern Markets*. Englewood Cliffs, New Jersey: Prentice-Hall.

Bem, Sandra. 1992. *The Lenses of Gender: Transforming the Debate on Sexual Inequality*. New Haven: Yale University Press.

Beneria, Lourdes. 1992. Accounting for Women's Work: The Progress of Two Decades. *World Development* 20(11): 1547–60.

Bernard, Jessie. 1981. *The Female World*. New York: Free Press.

Binkley, Marian. 1990. Work Organization among Nova Scotia Offshore Fisherman. *Human Organization* 49(4): 395–405.

Blumberg, Rea Lesser. 1984. A General Theory of Gender Stratification. In *Sociological Theory*, ed. Randall Collins. San Francisco: Jossey Bass.

Bort, John R. 1987. The Impact of Development on Panama's Small-Scale Fishermen. *Human Organization* 46 (3): 233–42.

Bottomore, T. B., ed. 1956. *Karl Marx: Selected Writing in Sociology and Social Philosophy*. New York: McGraw-Hill Book Company.

Buttel, Frederick N., and Howard Newby, eds. 1980. *The Rural Sociology of the Advanced Societies: Critical Perspectives*. Montclair, New Jersey: Allanheld, Osmun.

Caillouet, Charles W., Frank J. Patella, and William B. Jackson. 1980. Trends toward Decreasing Size of Brown Shrimp, Penaeus Aztecus, and White Shrimp, Penaeus Setiferus. In Reported Annual Catches from Texas and Louisiana, *Fishery Bulletin* 77(4): 985–89.

Calder, D. R., P. J. Eldridge, and E. B. Joseph, eds. 1974. The Shrimp Fishery of the Southeastern United States: A Management Profile. *Technical Report* 5 (September). Charleston: South Carolina Marine Resources Center.

Calhoun County Historical Commission. 1981. *The Shifting Sands of Calhoun County, Texas*. Calhoun County Historical Commission, Port Lavaca, Texas.

Canis, Wayne F. et al. 1985. *Living with the Alabama-Mississippi Shore*. Durham, North Carolina: Duke University Press.

Cannatella, Mary Michael, and Rita Emigh Arnold. 1985. *Plants of the Texas Shore*. College Station: Texas A & M University Press.

Center for Wetland Resources. 1979. Draft Environmental Impact Statement and Fishery Management Plan for the Shrimp Fishery of the Gulf of Mexico, United States Waters. Baton Rouge: Louisiana State University.

Christian, Richard Travis. 1986. Decision Making in Coastal Fisheries Conflict: The Case of Red Drum and Spotted Seatrout Legislation in Texas. Master's thesis, Texas A & M University.

Cicin-Sain, Biliana, Art Tiddens, and Cheryl Doss. 1985. Private Solutions to Conflicts over Public Resources: How Well Do They Work? Paper presented at the annual meeting of the American Fisheries Society, Sun Valley, Idaho.

Cicin-Sain, Biliana, Michael K. Orbach, Stephen J. Sellers, and Enrique Manzanilla. 1985. Conflictual Interdependence: United

States–Mexican Relations on Fishery Resources. Department of Political Science, University of California, Santa Barbara.

Clark, Joseph. 1955. *The Texas Gulf Coast: Its History and Development*. 2 vols. New York: Lewis Historical Publishing Company.

Clemenson, Brad. 1989. Judge Backs TEDs Suspension. *Mobile Press Register*, July, 29, 1A.

Coastal Society. 1990. Ports and Harbors: Our Link to the Water. *Proceedings of the Eleventh International Conference*, Bethesda, Maryland.

Coffman, Andrea G., ed. 1990. Bibliography for State Ocean Roles Panels. *Coastal Society Conference*, September.

Cole, Sally. 1991. *Women of the Praia*. Princeton: Princeton University Press.

Cox, Bruce. 1976. Statement by County Extension Marine Agent Cameron County before U.S. International Trade Commission. January 27. Brownsville, Texas.

Creighton, James. 1975. *A Narrative History of Brazoria County*. Brazoria, Texas: Brazoria County Historical Commission.

Crompton, John L., Dennis D. Beardsley, and Robert B. Ditton. 1976. Marinas on the Texas Gulf Coast. College Station: Texas Agricultural Experiment Station, Texas A & M University.

Crouch, Ben M., and Mark Miller. 1987. Lacey Act Enforcement in the Texas Gulf: A Sociological Analysis. In *Proceedings of the Conference on Gulf and South Atlantic Fisheries: Law and Policy*, ed. M. C. Jarman and D. K. Conners. Mississippi-Alabama Sea Grant Program 87-O13, New Orleans.

Curley, Stephen. 1990. Texas Coastal Plan: Analysis of a Failure. *Coastal Management* 18: 1–4.

———. ed. 1989. *Living on the Edge: Collected Essays on Coastal Texas*. Galveston: Texas A & M University.

Dalton, George, ed. 1967. *Tribal and Peasant Economies*. Garden City, New York: National History Press.

Danowski, Fran. 1980. Fishermen's Wives: Coping with an Extraordinary Occupation. *University of Rhode Island Marine Bulletin* 37. Narragansett: Marine Advisory Service, University of Rhode Island.

Danque, Neil C. N.d. The Causes of the Liability Insurance Crisis in Commercial Fishing. *May Day News*, Bellville, Texas.

Davis, Dona. 1986. Occupational Community and Fishermen's Wives in a Newfoundland Fishing Village. *Anthropological Quarterly* 59(3): 129–42.

_____. 1988. "Shore Skippers" and "Grass Windows": Active and Passive Roles for Women in a Newfoundland Fishery." In *To Work and To Weep: Women in Fishing Economies*, ed. Jane Nadel and Dona Davis. St. John's: Institute of Social and Economic Research, Memorial University of Newfoundland.

Dearmont, Lona. 1988. Coastal Barrier Resource Act of 1982: Bureaucratic Barriers. *Texas Shores* 21:16.

Denzin, Norman, K. 1989. *The Research Act: A Theoretical Introduction to Sociological Methods*. Englewood Cliffs, New Jersey: Prentice-Hall.

Ditton, Robert B., and Anthony J. Fedler. 1983. A Statewide Survey of Boatowners in Texas and Their Saltwater Fishing Activity. Sea Grant Program TAMU-SG-83-205, College Station: Texas A & M University.

Ditton, Robert B., Anthony J. Fedler, and Richard T. Christian. 1991. The Evolution of Recreational Fisheries Management in Texas.

Ditton, Robert B., Richard M. Harman, and Steve A. Woods. 1978. An Analysis of the Charter Boat Fishing Industry on the Texas Gulf Coast. *Marine Fisheries Review* (August): 1–7.

Donnelly, Marydele. 1990. National Academy of Sciences Recommends More TED Regulations. *Marine Conservation News* (Fall): 3.

_____. 1992. New TED Regulations Due Out after Months of Delay. *Marine Conservation News* (Spring): 15.

Donohue, John D. 1961. *Cam An: A Fishing Village in Central Vietnam*. East Lansing: Michigan State University Press.

Durrenberger, E. Paul. 1988. Shrimpers and Turtles on the Gulf Coast: The Formation of Fisheries Policy in the United States. *Maritime Anthropological Studies* 1, 2: 196–214.

_____. 1991. Psychology, Unions, and the Law: Folk Models and the History of Shrimpers' Unions in Mississippi. Paper presented at the annual meeting of the Society for Applied Anthropology, Charleston, South Carolina.

_____. 1992. *It's All Politics: South Alabama's Seafood Industry*. Urbana: University of Illinois Press.

Durrenberger, E. Paul, and Gisli Palsson. 1982. To Dream of Fish: The Causes of Icelandic Skippers' Fishing Success. *Journal of Anthropological Research* 38: 227–42.

Elkind, Peter. 1990. The Bucks Stop Here. *Texas Monthly* (January): 100–105.

Faludi, Susan. 1991. *Backlash: The Undeclared War against American Women*. New York: Crown Publishers.

Field, Donald W. et al. 1991. *Coastal Wetlands of the United States: An*

Accounting of A Valuable National Resource. National Oceanographic and Atmospheric Administration. Rockville, Maryland: GPO.

Fikac, Peggy. 1985. Shrimp, Oyster Rules Will Be under Wildlife Department. *Brownsville Herald*, April 30, 9.

Finnan, Christine R. 1982. Community Influences on the Occupational Adaptation of Vietnamese Refugees. *Anthropological Quarterly* 55(3): 161–69.

Fiske, Shirley J., and Francis M. Schuler. 1990. Response to Change: Perspective from the Social Sciences. *Ocean and Shoreline Management* 13(304): 167–78.

Flint, R. Warren. 1983. Computer-Accessible Annotated Bibliography of the Corpus Christi Bay Estuary. Sea Grant College Program TAMU-SG-83-605, NA81AA-D00092, R/ES-1. College Station: Texas A & M University Press.

Flint, R. Warren, and Nancy N. Rabalais. 1981. Gulf of Mexico Shrimp Production: A Food Web Hypothesis. *Fishery Bulletin* 79(4): 737–48.

Flippin, Louise. 1979. The Indochinese Refugee Situation: America's Newest Wave of Immigrants. Center for Asian Studies. Austin: University of Texas.

Fricke, Peter. 1989. The Use of Sociological Data in the Allocation of Common Property Resources: A Comparison of Practices. Paper presented at the annual meeting of the American Fisheries Society, Ithaca, New York, August.

Friend, J. H. et al. 1981. A Socioeconomic Study. Alabama Coastal Region Ecological Characterization FWS/OBS-81/41, vol. 3:367. U.S. Fish and Wildlife Service, Office of Biological Services, Washington, D.C.

Garfield, Nina. 1991. Perceptions of the Texas State and Federal Closures among Inshore Shrimpers in Galveston. Master's thesis, University of Rhode Island.

Gatewood, John B., and Bonnie J. McCay. 1988. Job Satisfaction and the Culture of Fishing: A Comparison of Six New Jersey Fisheries. *Maritime Anthropological Studies* 1: 103–28.

_____. 1990. Comparison of Job Satisfaction in Six New Jersey Fisheries: Implications for Management. *Human Organization* 49(1): 14–15.

Geertz, Clifford. 1963. *Peddlers and Princes.* Chicago: University of Chicago Press.

Gersuny, Carl, and John J. Poggie, Jr. 1973. The Uncertain Future of Fishing Families. *Family Coordinator* (April): 241–44.

Gillespie, William C., James C. Hite, and John S. Lytle. 1969. An

Econometric Analysis of the U.S. Shrimp Industry. *Economics of Marine Resources* 2. Department of Agricultural Economics and Rural Sociology, South Carolina Agricultural Experiment Station, Clemson University.

Gillette, Michael L. 1986. *Texas in Transition*. Austin: Lyndon B. Johnson School of Public Affairs, Lyndon Baines Johnson Library, University of Texas.

Gray, Edith, and Richard Jensen. 1988. Shrimping Industry vs. Feds. *Baldwin Today*, May 17, 1C.

Greer, Germaine. 1991. *The Change: Women, Aging, and Menopause*. New York: Alfred A. Knopf.

Griffith, David C. 1987. Nonmarket Labor Processes in an Advanced Capitalist Economy. *American Anthropologist* 89(4): 838–52.

Guest, William C. 1958. The Texas Shrimp Fishery. *Bulletin* 36. Marine Laboratory, Texas Game and Fish Commission, Austin.

Gunter, Valerie J. 1983. Environmental Conflicts in the Coastal Zone, typescript. Texas A & M University.

Gunther, Nancy C., and B. G. Gunther. 1990. Domestic Division of Labor among Working Couples: Does Androgyny Make a Difference? *Psychology of Women Quarterly* 14: 355–70.

Guthrie, Keith. 1988. *Texas Forgotten Ports*. Austin: Eakin Press.

Haney, Wava G., and Jane B. Knowles. 1988. *Women and Farming: Changing Roles, Changing Structures*. Boulder, Colorado: Westview Press.

Harrigan, Stephen. 1980. *Aransas*. New York: Alfred A. Knopf.

Harris, Craig K. 1986. Toward A Sociology of Fisheries. *WHOI Proceedings*. Woods Hole Oceanographic Institute, Woods Hole, Massachusetts.

Heilbroner, Robert L. 1962. *The Making of Economic Society*. Englewood Cliffs, New Jersey: Prentice-Hall.

Henry Walter K., Dennis M. Driscoll, and J. Patrick McCormack. 1983. Hurricanes on the Texas Coast. Sea Grant College Program TAMU-SG-75-504. College Station: Texas A & M University Press.

Henwood, Tyrrell A., and Warren E. Stuntz. 1987. Analysis of Sea Turtle Captures and Mortalities during Commercial Shrimp Traveling. Fish and Wildlife Service, Washington, D.C. *Fishery Bulletin* 85 (4): 813–17.

Herskovits, Melville J. 1952. *Economic Anthropology*. New York: W. W. Norton.

Hester, Thomas R. 1980. A Chronological Overview of Prehistoric Southern and South-Central Texas. In *Prehistory of Northeastern*

Mexico and Adjacent Texas, ed. Jonathan Epstein, Thomas R. Hester, and Herbert Graves. Center for Archaeological Research, San Antonio: University of Texas Press.

Hickey, Gerald C. 1969. *Village in Vietnam*. New Haven: Yale University Press.

Hobson, J. A. 1971. *Imperialism*. Ann Arbor: University of Michigan Press.

Hochschild, Arlie, with Anne Machung. 1989. *The Second Shift: Working Parents and the Revolution at Home*. New York: Viking Press.

Hoese, H. Dickson, and Richard H. Moore. 1977. *Fishes of the Gulf of Mexico: Texas, Louisiana and Adjacent Waters*. College Station: Texas A & M University Press.

Hollin, Dewayne, and Malon Scogin, eds. 1981. Directory of Texas Shipyards. *Sea Grant Publication* SG-80-507. College Station: Texas A & M University Press.

Jedlicka, Davor. 1981. A Demographic Study of the Coastal Counties of Georgia, 1790–1980. Georgia Marine Science Center, University System of Georgia. *Technical Report Series* (January): 81–83.

Jedlicka, Davor, and Ray-May Hsun. 1980. Coastal Farming and Fishing. Paper presented at the fifth World Congress for Rural Sociology, Mexico City.

Jensen, Joan M. 1986. *Loosening the Bonds: Mid-Atlantic Farm Women, 1750–1850*. New Haven: Yale University Press.

Johnson, Fred F., and Milton J. Linder. 1934. Shrimp Industry of the South Atlantic and Gulf States. *Investigative Report* 21. Bureau of Fisheries, Washington, D.C.

Johnson, Jeffrey C., and Michael K. Orbach. 1990. Migratory Fishermen: A Case Study in Interjurisdictional Natural Resource Management. *Ocean and Shoreline Management* 13 (304): 231–52.

Johnson, Ronald N., and Gary D. Libecap. Regulatory Arrangements and Fishing Rights in the Texas Bay Shrimp Fishery. Typescript. Department of Economics, Texas A & M University.

Jones, Lonnie L., John W. Adams, Wade L. Griffin, and Jeffrey Allen. 1974. Impact of Commercial Shrimp Landings on the Economy of Texas and Coastal Regions. *Sea Grant Publication* TAMU-SG-75-204. Agricultural Experiment Station, College Station: Texas A & M University.

Juster, Norton. 1979. *So Sweet to Labor: Rural Women in America, 1865–1895*. New York: Viking Press.

Kaiser, Ronald A. 1987. *Hand Book of Texas Water Law: Problems and*

Needs. Texas Water Resources Institute, Agricultural Experiment
Station, College Station: Texas A & M University.

Kanamine, Linda. 1992. Report: Toxic Emissions Down but They Still
Pose a Health Risk. *USA Today,* October 1, 13A.

Keithly, Walter R., Jr., et al. 1992. Effects of Shrimp Aquaculture on the
U.S. Market: An Econometric Analysis. Typescript. Center for
Wetland Resources, Louisiana State University, Baton Rouge.

Kelly, Gail Paradise. 1977. *From Vietnam to America.* Boulder, Colorado:
Westview Press.

Klima, Edward F., et al., eds. 1987. Review of the 1986 Texas Closure
for the Shrimp Fishery of Texas and Louisiana. NOAA Technical
Memorandum NMFS-SEFC-197 (July). Galveston Laboratory,
Southeast Fisheries Center, National Marine Fisheries Service,
NOAA, U.S. Department of Commerce.

Kondo, Dorinne E. 1990. *Crafting Selves: Power, Gender, and Discourses of
Identity in a Japanese Workplace.* Chicago: University of Chicago
Press.

Kutkuhn, Joseph H. 1962. Gulf of Mexico Commercial Shrimp
Populations: Trends and Characteristics, 1956–1959. *Fishery Bulletin*
212 (62): 307–12.

Laurent, Eugene A. 1974. Description of Industry: Economics. In *The
Shrimp Fishery of the Southeastern United States: A Management
Planning Profile,* ed. D. R. Calder. Technical Report 5: 70–85. South
Carolina Marine Resources Center, Charleston.

Lawrence, Addison L., George W. Chamberlain, and David L.
Hutchins. 1981. Shrimp Mariculture. *Sea Grant Publication*
TAMU-SG-82-503. College Station: Texas A & M University Press.

Leap, William L. 1977. Maritime Subsistence in Anthropological
Perspective: A Statement of Priorities. In *Those Who Live from the
Sea,* ed. Estellie Smith, American Ethnological Society. St. Paul,
Minnesota: West Publishing Company.

Levine, Edward B., and Bonnie J. McCay. 1987. Technological Adoption
among Cape May Fishermen. *Human Organization* 46 (3): 243–53.

Lin, K., M. Masuda, and L. Tazuma. 1979. Vietnamese Refugees: A
Three-Year Longitudinal Study. *Psychiatric Journal of the University of
Ottawa* 9: 79–84.

Linder, Milton J., and William W. Anderson. 1956. Growth, Migration,
Spawning, and Size Distribution of Shrimp, Penaeus Setiferus.
Fish and Wildlife Service. *Fishery Bulletin* 106 (56): 555.

Liu, William, Mary Anne Lamanna, and Alice Murata. 1979. *Transition to Nowhere: Vietnamese Refugees in America*. Nashville, Tennessee: Charter House.

Lopata, Helene Z. 1993. The Interweave of Public and Private: Women's Challenge to American Society. *Journal of Marriage and the Family* 55: 176–90.

Macinko, Seth. 1991. Gospels of Efficiency, Public Policy and Rural Poverty: The Case of the U.S. Commercial Fishing Industry. Paper presented at the annual meeting of the Rural Sociology Society, Columbus, Ohio, 1991.

Maril, Andrea Fisher. 1993. Women and Shrimping: The Texas Way. Paper presented at the annual meeting of Society for Applied Anthropology. San Antonio, Texas.

Maril, Robert Lee. 1979. Shrimping in South Texas: Social and Economic Marginality Fishing for a Luxury Commodity. Paper presented at the annual meeting of the Association for Humanist Sociology, Johnstown, Pennsylvania.

————. 1982. Continuity and Change in a Texas Fishing Community. Paper presented at the annual meeting of the American Anthropological Association, Washington, D.C., December.

————. 1983a. *Texas Shrimpers: Community, Capitalism, and the Sea*. College Station: Texas A & M University Press.

————. 1983b. Condos and the Poor: The Texas Coast in the 1990s. Typescript. Texas Southmost College.

————. 1984. They Followed Us with Jets and My Own Country Arrested Us: The Arbitrary Enforcement of the Lacey Act. Paper presented at the annual meeting of the Rural Sociological Society, College Station, Texas.

————. 1986. *Cannibals and Condos: Texans and Texas along the Gulf Coast*. College Station: Texas A & M University Press.

————. 1989. *Poorest of Americans: The Mexican Americans of the Lower Rio Grande Valley of Texas*. Notre Dame, Ind.: University of Notre Dame Press.

Maril, Robert Lee, and Anthony N. Zavaleta. 1979. Drinking Patterns of Low-Income Mexican-American Women. *Journal of Studies on Alcohol* 40(5): 480–85.

Marr, John Columbus. 1928. *The History of Matagorda County, Texas*. Master's thesis, University of Texas.

Martin, Norman. 1988. Living in the Coastal Zone. *Texas Shores* 21(2): 4–8.

Marx, Karl. 1964. *Pre-Capitalist Economic Foundation*. New York: International Publishers.

Matagorda County Historical Commission. 1986. *Historic Matagorda County*. 2 vols. Houston: O. Armstrong Company.

Mauro, Garry. 1990. Address to the Coastal Society International Conference. Coastal Society *Bulletin* 13(4): 14–16.

McGoodwin, James Russell. 1980. The Human Costs of Development. *Environment* 22(1): 25–42.

———. 1987. Mexico's Conflictual Inshore Pacific Fisheries: Problem Analysis and Policy Recommendation. *Human Organization* 46(3): 221–32.

Mearns, Anna. 1992. Adventures of Ranger Rick. *Ranger Rick* (June): 16–20.

Meltzoff, Sarah Keene, and Edward Lipuma. 1985/1986. The Social Economy of Coastal Resources: Shrimp Mariculture in Ecuador. *Culture and Agriculture* 28 (Winter): 1–18.

Miller, Charlotte, L., and John P. Nichols. 1985. Economics of Harvesting and Market Potential for the Texas Blue Crab Industry. *Sea Grant Publication* TAMU-SG-86-201. Department of Agricultural Economics, College Station: Texas A & M University Press.

Miller, Michael V., and William P. Kuvlesky. 1975. The Farm Labor Movement in South Texas: Historical Development, Current Status, and Implications for Change. Department of Rural Society, College Station, Texas.

Moffett, A. W. 1967. *The Shrimp Fishery in Texas*. Austin: Texas Parks and Wildlife Department.

Montero, Darrel. 1979. *Vietnamese Americans: Patterns of Resettlement and Socioeconomic Adaptation in the United States*. Boulder, Colorado: Westview Press.

Morris, Jim. 1992. Shrimpers Cut Loose Again. *Houston Chronicle*, January 5, 1C.

Morris, Julie, and Mark Mayfield. 1990. More and More Risk for the Gulf. *USA Today*, June 12, 1A.

Mountain, Karen et al. 1984. The Rural Texas Environment: A Profile of Stressors. Texas Rural Health Field Services Program (September). Center for Social Work Research, School of Social Work, University of Texas.

Mullarkey, Nora E., Mary Walker, and Yvonne D. Knudson. 1984. Water and Rural Texas: A Resource Guide. Texas Rural Health Field Services Program. Center for Social Work Research, School of Social Work, University of Texas.

Mullen, Patrick B. 1969. The Function of Magic Folk Belief among Texas Coastal Fishermen. *Journal of American Folklore* 82: 214–25.

_____. 1978. *I Heard the Old Fishermen Say: Folklore of the Texas Gulf Coast*. Austin: University of Texas Press.

Murray, Laura. 1989a. Industry Promotion. *Texas Shores* 21, 4.

_____. 1989b. New Horizons. *Texas Shores* (Summer–Fall): 18–20.

Nadel, Jane, and Dona Davis. 1988. Introduction. In *To Work and to Weep: Women in Fishing Economies*, ed. Jane Nadel and Dona Davis. Institute of Social and Economic Research, St. John's: Memorial University of Newfoundland.

Nance, James M., and Edward F. Klima et al. 1988. Review of the 1987 Texas Closure for the Shrimp Fishery off Texas and Louisiana. NOAA Technical Memorandum SEFC-NMFS (February). Galveston Laboratory, National Marine Fisheries Center, National Marine Fisheries Service, NOAA, U.S. Department of Commerce.

National Oceanographic and Atmospheric Administration. 1987. *Draft Supplement to the Final Environmental Impact Statement on TEDs*. National Marine Fisheries Service, U.S. Department of Commerce. Washington, D.C.: GPO.

_____. 1990. *Estuaries of the United States: Vital Statistics of a National Resource Base*. Rockville, Maryland: GPO.

Natural Resources Division. N.d. *Coastal Natural Resources in Texas: A Report to the Governor and the 68th Legislature*. Texas Energy and Natural Resources Advisory Council, Austin.

Nelson, Robert G., and William E. Hardy, Jr. 1980. *The Economic and Environmental Structure of Alabama's Coastal Region, Part 1: Economic Structure* MASGP-79-016. Department of Agricultural Economics and Rural Sociology. Agricultural Experiment Station, Auburn University.

Nichols, John P. et al. 1980. Marketing Alternatives for Fishermen. *Sea Grant Publication* TAMU-SG-1-80-204. College Station: Texas A & M University Press.

Nix, Harold L., and Muncho, Kim. 1982. *A Sociological Analysis of Georgia Commercial Shrimp Fishermen, 1976–77*. Institute for Community and Area Development, University of Georgia.

O'Neil, P. E. et al. 1982. Coastal Bibliography. *Alabama Coastal Region Ecological Characterization* FWS/OBS-82/21, vol. 1:404. U.S. Fish and Wildlife Service, Office of Biological Services, Washington, D.C.

_____. 1982. A Synthesis of Environmental Data. *Alabama Coastal Region Ecological Characterization* FWS/OBS-82/42, vol. 2:346. U.S. Fish and Wildlife Service, Office of Biological Services, Washington, D.C.

Orbach, Michael K. 1977. *Hunters, Seamen, and Entrepreneurs: The Tuna Seinermen of San Diego*. Berkeley: University of California Press.

Ormerord, Leonard. 1957. *The Curving Shore*. New York: Harper and Brothers.

Orth, Frank et al. 1981. Minority Participation in the Fisheries of the Gulf and South Atlantic. Report prepared for the Gulf and South Atlantic Fisheries Development Foundation, Tampa, Florida.

Ortner, Sherry B., and Harriet Whitehead. 1981. Introduction: Accounting for Sexual Meanings. In *Sexual Meanings: The Cultural Construction of Gender and Sexuality*, ed. Sherry B. Ortner and Harriet Whitehead. Cambridge: Cambridge University Press.

Palmer, Craig T. 1990. Telling the Truth (Up to a Point): Radio Communication among Marine Lobstermen. *Human Organization* 49(2): 157–63.

Parker, A. Lelland. 1983. *The Coastal Bend: A Pictorial History*. Corpus Christi, Texas: South Coast Publishing Company.

Perez, Lisandro. 1979. Working Offshore: A Preliminary Analysis of Social Factors Associated with Safety in the Offshore Workplace. *Sea Grant Publication* LSU-T-79-001, Center for Wetland Resources, Louisiana State University.

Pesson, L. L. 1974. *The Coastal Fishermen of Louisiana: Their Characteristics, Attitudes, Practices, and Responses to Change*. Cooperative Extension Service, Center for Agricultural Sciences and Rural Development, Louisiana State University.

Peterson, Susan, and Leah Smith. 1979. New England Fishing, Processing, and Distribution. *Technical Report* WHOI-79-52 (March). Woods Hole Oceanographic Institute, Woods Hole, Massachusetts.

Peterson, Susan, and David Georgianna. 1988. New Bedford's Fish Auction: A Study in Auction Method and Market Power. *Human Organization* 47(3): 43–52.

Poggie, John J., Jr., and C. Gersuny. 1974. Fishermen of Galilee. *Marine Bulletin* 17. Rhode Island Sea Grant Program.

Pollnac, Richard B., and John J. Poggie, Jr. 1988. The Structure of Job Satisfaction among New England Fishermen and Its Application to Fisheries Management Policy. *American Anthropologist* 90: 888–901.

Pomeroy, Carolina. 1989. Half Moon Caye Natural Monument. Master's thesis, University of Miami.

Prochaska, Fred J., and James C. Cato. 1977. An Economic Profile of Florida Commercial Fishing Firms: Fishermen, Commercial Activities, and Financial Considerations. *State University System of Florida Sea Grant Program Report* 19 (February). Food and Resource

Economics Department, Institute of Food and Agricultural Sciences, University of Florida.

Prochaska, Fred J., Mauro Suazo, and Walter R. Keithly. 1983. World Shrimp Production Trends and the U.S. Import Market. In *Proceedings of the Eighth Annual Tropical and Subtropical Fisheries Conference of the Americas.* Sea Grant College Program TAMU-SG-112 (August). College Station: Texas A & M University Press.

Radcliffe, Lewis. 1920. Report Appendix X: Fishery Industries of the U.S. *Report of the Division of Statistics and Methods of the Fishery for 1919,* Document 892: 180–89. Bureau of Fisheries, Washington, D.C.

Reissman, Catherine Kohler. 1990. *Divorce Talk: Women and Men Make Sense of Personal Relationships.* New Brunswick, New Jersey: Rutgers University Press.

Reskin, Barbara, and Irene Padavic.1994. *Women and Men at Work.* Thousand Oaks, Maryland: Pine Forge Press.

Reyer, Anthony J. et al. 1988. *The Distribution and Areal Extent of Coastal Wetlands in Estuaries of the Gulf of Mexico.* Rockville, Maryland: National Oceanographic and Atmospheric Administration.

Riley, Glenda. 1988. *The Female Frontier: Comparative View of Women on the Prairie and the Plains.* Lawrence: University Press of Kansas.

————. 1992. *A Place to Grow: Women in the American West.* Arlington Heights, Illinois: Harland Davidson.

Ritter, Mary Christine. 1991. Sport Fishery Management in East Matagorda Bay (Texas): An Analysis of Decision Making. Master's thesis, Texas A & M University.

Roberts, K. J., and M. E. Sass. 1979. Financial Aspects of Louisiana Shrimp Vessels, 1978. *Sea Grant Publication* LSU-TL-79-007 (December). Center for Wetlands Resources, Louisiana State University.

Rosenfeld, Rachel Ann. 1985. *Farm Women: Work, Farm and Family in the United States.* Chapel Hill: University of North Carolina Press.

Rubin, Lillian Breslow. 1976. *Worlds of Pain: Life in the Working-Class Family.* New York: Basic Books.

Rudloe, Jack, and Anne Rudloe, 1989. Shrimpers and Lawmakers Collide over a Move to Save the Sea Turtles. *Smithsonian* (December): 45–54.

Russell, Susan D., and Maritsa Poopetch. 1990. Petty Commodity Fishermen in the Inner Gulf of Thailand. *Human Organization* 49(2): 174–87.

Sachs, Carolyn E. 1983. *The Invisible Farmers: Women in Agricultural Production.* Totowa, New Jersey: Rowman and Allanheld.

———. 1993. Women Now Identify Themselves as Farmers, Not Helpers. *Stillwater News Press,* February 7, D1.

Salinas, Martin. 1990. *Indians of the Rio Grande Delta.* Austin: University of Texas Press.

Santopietro, George D., and Leonard A. Shabman. 1985. Evaluating Public Policy Options for Improving Water Quality and Increasing Oyster Production in the Chesapeake Bay. Paper presented at the meeting of American Fisheries Society, Sun Valley, Idaho.

Sass, M. E., and K. S. Roberts. 1979. Characteristics of the Louisiana Shrimp Fleet, 1978. *Sea Grant Publication* LSU-TL-79-006 (December). Center for Wetland Resources, Louisiana State University.

Schwartz, Felice. 1992. *Breaking with Tradition: Women and Work, the New Facts of Life.* New York: Warner Books.

Sea Grant Publications. 1976. Sea Grant Publications, 1969–75. *Sea Grant Publication* TAMU-SG-76-604. Sea Grant College Program, College Station: Texas A & M University Press.

Shipp, Robert L. 1986. *Fishes of the Gulf of Mexico.* Mobile, Alabama: 20th Century Publishing Company.

Shostak, Arthur B. 1980. *Blue Collar Stress.* Reading, Massachusetts: Addison-Wesley Publishing Company.

Shrimpers Face Impending TED Laws, Extended Closure Units. 1989. *Texas Shoreline* (March): 2–3.

Shrimpers Win Reprieve from Turtle Device Laws. 1988. *Texas Shoreline* (June): 1–3.

Smith, M. Estellie, ed. 1977. *Those Who Live from the Sea.* American Ethnological Society. St. Paul, Minnesota: West Publishing Company.

Soden, Dennis L. 1988. Community-Oriented Analysis: Viewing Policy Alternatives for the Gulf Coast Region. Working paper. In Marine Resource Utilization: A Conference on Social Science Issues, ed. J. Stephen Thomas, Robert Lee Maril, and E. Paul Durrenberger. University of South Alabama and the Mississippi Sea Grant Consortium, Mobile, 1989.

Southern Exposure. 1982. *Southern Exposure* 10, 3 (May–June).

Spruce, William E. 1985. Indochinese Refugees on the Central Coast of Texas. Typescript. Foreign Service Institute, U.S. Department of State.

Starr, Paul D. 1979. Troubled Waters: Vietnamese Fisherfolk on America's Gulf Coast. *International Migration Review* 13: 25–45.

Stein, Barry N. 1979. Occupational Adjustment of Refugees: The Vietnamese in the United States. *International Migration Review* 1: 25–45.

Stevenson, Charles H. 1893. Report on the Coastal Fisheries of Texas. *Report of the Commissioner for 1889–91, 17:373–420. U.S. Commission of Fish and Fisheries,* Washington, D.C.

Stewart, Sharon. 1984. Resource Use and Use Conflicts in the Exclusive Economic Zone. In *The Texas Initiative,* ed. Lynne Carter Hanson and Carol Alexander. Wakefield, Rhode Island: Time Press.

Stickney, Robert R., and James T. Davis. 1981. Aquaculture in Texas: A Status Report and Development Plan. *Sea Grant Publication* TAMU-SG-81-119 (August). Sea Grant College Program, College Station: Texas A & M University Press.

TEDs Agreement Prompts Senator to Drop Protests. *Brazosport Facts,* July 15, 1988.

Tettey, Ernest, Christopher Pardy, Wade Griffin, and A. Nelson Swartz. 1984. Implications of Investing under Different Economic Conditions on the Profitability of Gulf of Mexico Shrimp Vessels Operating out of Texas. *Fishery Bulletin* 82(2): 385–94.

Texas Coast Hurricanes. 1986. *Texas Coast Hurricanes* TAMU-SG-86-505, NA85AA-D-SG128 (May). College Station: Texas A & M University.

Texas Coastal and Marine Council. 1982. *Texas Coastal Legislation,* 5th ed. Austin: Texas Coastal and Marine Council.

_____. 1983. *Texas Bay Shrimp Industry: Status Report and Recommendations* (June). Austin: Texas Coastal and Marine Council.

_____. 1987. *Hearings on Bay Shrimpers.* Austin: Texas Coastal and Marine Council.

Texas Energy and Natural Resources Policy Research Project. 1980. Texas Coastal Zone Issues. *Policy Research Project Report.* Lyndon B. Johnson School of Public Affairs, University of Texas.

Texas Parks and Wildlife Department. 1988. *A Guide to Commercial Fishing Regulations* PWD-BK-3400, 74-8/88. Austin: Fisheries and Wildlife Division.

_____. 1991a. *Trends in Texas Commercial Fishing Landings, 1972–1990.* Austin: Fisheries and Wildlife Division.

_____. 1991b. *The Texas Shrimp Fishery: A Report to the Governor and the 72nd Legislature.* Austin: Fisheries and Wildlife Division.

_____. 1991c. *Meeting of Shrimp Advisory Committee* (June 26). Austin: Fisheries and Wildlife Division.

_____. 1991d. *Transcript of Meeting of Shrimp Advisory Committee* (December 2). Austin: Fisheries and Wildlife Division.

Thompson, Earl. 1976. *Caldo Largo.* New York: New American Library.

Trans-Century Corporation. 1979. New Economic and Social Opportunities for Americans and Indochinese on the Texas Gulf Coast. Report to Minority Business Development Agency. U.S. Department of Commerce, Washington, D.C.

Tveten, John L. 1982. *Coastal Texas.* College Station: Texas A & M University Press.

U.S. Congress. 1990. Office of Technology Assessment. *Coping with an Oiled Sea* OTA-BP-O-63. Washington, D.C.: U.S. GPO.

U.S. International Trade Commission. 1985. Conditions of Competition Affecting the U.S. Gulf and South Atlantic Shrimp Industry. Washington, D.C.: U.S. GPO.

Vondruska, John. 1987. *The Gulf Shrimp Industry.* St. Petersburg, Florida: National Marine Fisheries Service.

_____. 1991. World Shrimp Situation 1990: Effects on Southeast Harvesters. *NOAA Technical Memorandum* NMFS-SEFC-294. Galveston Laboratory, Southeast Fisheries Center, National Marine Fisheries Service, NOAA, U.S. Department of Commerce.

_____. 1992. Southeast Shrimp Fishery Market Conditions, 1991–1992. Preliminary draft. St. Petersburg, Florida: National Marine Fisheries Service.

Wadel, Cato. 1973. Now, Whose Fault Is That? *Newfoundland Social and Economic Studies* 11. Institute of Social and Economic Research, Memorial University of Newfoundland. Toronto: University of Toronto Press.

White, David R. M. 1977. Environment, Technology, and Time-Use Patterns in the Gulf Coast Shrimp Fishery. In *Those Who Live from the Sea,* ed. M. Estellie Smith. American Ethnological Society. St. Paul, Minnesota: West Publishing Company.

_____. 1989. Knocking 'em Dead: The Formation and Dispersal of Work Fleets among Alabama Shrimp Boats. *Maritime Anthropological Studies* 2, 1: 69–79.

_____. 1991. Skipper Effect or Fleet Effect? How Alabama Shrimpers Find Shrimp. Paper presented at the annual meeting of the Society for Applied Anthropology. Charleston, South Carolina.

Woolley, Bryan, and Skeeter Hagler. 1985. *Where Texas Meets the Sea.* Dallas: Pressworks.

INDEX

❖   ❖   ❖